CLIMATE CHANGE
SERVICE ECONOMY
AND INDUSTRY

With Empirical Cases & Theories

To my Lord, Jesus Christ

CLIMATE CHANGE, SERVICE ECONOMY AND INDUSTRY

With Empirical Cases & Theories

JUNMO KIM

Konkuk University, Seoul, Korea

iUniverse, Inc.

New York   Bloomington

CLIMATE CHANGE, SERVICE ECONOMY AND INDUSTRY

With Empirical Cases & Theories

iUniverse books may be ordered through booksellers or by contacting:

iUniverse
1663 Liberty Drive
Bloomington, IN 47403
www.iuniverse.com
1-800-Authors (1-800-288-4677)

ISBN: 978-1-4502-6474-7 (pbk)
ISBN: 978-1-4502-6475-4 (ebk)

Printed in the United States of America

iUniverse rev. date: 10/20/10

Contents

List of Figures

List of Tables

Preface

Climate and weather have seemed to be exogenous variables to economy and industry. Or at least beneath the surface, it has been regarded in this way. Against this widely accepted tradition, more and more audience and forerunners in science, meteorology, consulting and academic circles have become aware of the fact that weather and climate do affect our lives, industry, and economy.

Taking this as a backdrop, this book covers the issues of how economy and industry perceives weather and climate as something that would affect their destiny. As always for the development of discourse in many other fields, this book shares some notion with earlier writings on climate change, while differs from the earlier ones on the point that this book tried to capture industry side ideas. Usually books on climate change tend either bring scientific knowledge or advocacy or policy ideas from policy circles that could be borrowing strengths from science community. In some cases, books had critical reasoning that tried to warn against blinded following of climate change as a trend. This book tried, instead, to reflect what industries think and how they think they are preparing based on empirical surveys.

In terms of approach, this book took an unique one. That is while surveys utilized in the book were drawn from the case of Korea, always finding implications were aimed at broader scope so that other national cases could be fit into the frame work. For example, in presenting future items to be introduced, the Korean case worked as an example, through which other societies can easily find compatibility.

Turning into the motivation to write this book, it is possible to mention my almost 10 year consulting experiences for the meteorology administration of Korea and other private weather service providers in Korea since 1998. With the long haul experience, I felt that it is a time to sum up what has been experienced. Yet, this book is too limited to show everything that has been discussed and written in policy circles. Contents in the book are a partial introduction of the whole story. Also it is noteworthy that with the sabattical research year in 2009, writing of this book was made possible. The publication of this book clearly offers an invaluable milestone in my research after publishing *The South Korean Economy* (Ashgate 2002), *Globalization and Industrial Development* (iUniverse 2005), and *R&D and Economy*(Iuniverse 2005).

Upon publishing this book, I would like to express deep gratitude to my teachers at all levels of my schooling. Above all, I would like to express my deepest gratitude to my Lord Jesus Christ for allowing inspirations for research. Though not mentioned, I believe that there are many other significant persons who have given positive encouragements to my life and research. Finally, I would like to express deep gratitude to my wife Hyeree and my four kids, Gyu-Young, Ah-Young, Je-Hyun, and Je-Yoon, who have created & expanded externalities of joy in life.

CHAPTER 1
INTRODUCTION

Prologue: Why organizing technology development matters?: Technology Aspect

1. Transition from Manufacturing dominated to Service economy

As economies move from a traditional economy to a more knowledge intensive one (Drucker 2002; Baily 2001; Kim, J. 2008), the importance of service sector has been growing in every sphere of world where past industrialization has reached at a substantial level (Chandler 1990; Kim, J. 2001a; Kim, J. 2006a). Amongst this change, our global economy has been also 'forced' to adapt to a new surrounding, namely 'Climate Change' (Dryzek 2005; drake 2000; Emanuel 2007).

When one thinks of climate change, it usually denotes a huge concept, and therefore it seems that it has little implication to our daily lives. Yet, when one thinks of climate change and emphasis on weather information from a tradition of traditional economy into a service oriented knowledge economy, it would be more meaningful.

Transiting toward knowledge economy, in which service sector and its affected manufacturing sectors dominate, factors that have not been considered as factors of production have begun regarded as key production factors (Kim, J. 2009). One of the clear examples would be the factor of weather or meteorology (Kim, J. 2003, 2005,2007, 2008). In fact, virtually all manufacturing and service sectors are affected by weather information in its profitability. Climate is a longer term version of weather, usually with a time frame of several months. In this respect, climate change, even without the context of climate change, is an important factor in both manufacturing and service sectors (Zilman 2001). Add to this, dynamic change of climate will definitely increase complexity with which economies would face (Emanuel 2007; Dryzek 2005; Grubb & Wilde 2004).

To discuss the contexts that would increase the importance of meteorology and climate change, it may be possible to list several backgrounds.

Changing industrial landscape

Beyond once popular notion of the post-industrial society, our economy has been undergoing a structural change in which manufacturing and service sectors are converging into a new being (Kim, J. 2008). In today's contexts, advanced manufacturing cannot exist without help from service sectors and vice versa (Hansen 2002; Kim, J. 2005; Dunning 2005). Among the influential elements that can glue the two sectors exist weather and climate information. Even among service sectors, there is dependence between service sectors. For example, finance sector rely heavily on weather for such transactions like derivatives, which include future options trade in agricultural products or natural resources like petroleum.

Information Era

Also going beyond the already established notion of information society and age, our global economy is speeding up into the information era, which transforms so many types of knowledge into tradable knowledge. Once only academically regarded, DNA or gene info is clearly becoming a tremendous source of economic opportunities. Likewise, weather and climate information have potential to be treated as a high value enhancing source for economic activities (Fleming 2004; Flannery 2006).

Diversity

One of the key phenomenon to describe 20th and 21st century in common would be increased diversity across diverse areas. This would very naturally increase the demand for specific service and goods, which, in turn, would demand more specific information. Weather information, together with finance information, can be examples. To describe a little further, different life styles would request more specific weather forecasts for different purposes. Add to this, increased degrees of climate change would segmentize potential markets (Nordhaus 2003).

The contexts of the Korean economy

After reviewing general contexts, it is appropriate to focus on the Korean contexts. Manufacturing sectors, the main growth engine of Korea, have contributed to the country's economic development with an average growth of 11.2% during 1971 to 1997 period. Especially during the 1971 to 1987 period, the country's manufacturing marked an average rate of 14.3% growth. In contrast, during 1988-1997 period, the country showed about 8.1% growth in manufacturing, which is a sharp decline compared to previous periods (Kim, J. 2002, 2005).

From the table in the below, it is possible to infer several key changes that have occurred in the Korean economy.

First, manufacturing sector was the locomotive in producing high speed growth of the economy, which can be inferred from the higher annual increase rate of growth in manufacturing vis-à-vis that of the real GDP. The fact that manufacturing had an important role can also be evidenced by the figures of labor productivity and growth rate of Total fixed capital formation. The key contribution of the manufacturing sector began declining from the late 1980s. The gap between the growth rate of manufacturing and that of economy as a whole has been reduced significantly (Sakong 1987).

		1971-97	71-87	88-97	71-80	81-87	1988-97
All Sctors	Real GDPgrowth rate	7.8	8.1	7.5	7.5	8.7	7.5
	Growth rate of Total fixed capital formation	10.9	11.6	10.2	13.7	9.5	10.2
	Growth rate of fixed investment	13.0	15.8	10.2	20.8	10.8	10.2
	Growth rate of private consumption	6.9	6.2	7.6	5.2	7.2	7.6
	Export growth rate	15.0	18.0	11.9	21.6	14.3	11.9
	Inflation	10.8	14.4	7.2	21.3	7.4	7.2
	unemployment	3.3	4.1	2.4	4.1	4.0	2.4
Manu-facturing	Labor Productivity(Y/L) Growth rate	5.7	5.8	5.5	4.9	7.1	5.5
	Growth rate	11.2	14.3	8.1	15.9	12.7	8.1
	Growth rate of Total fixed capital formation	15.0	19.9	10.0	22.8	16.9	10.0
	Growth rate of fixed capital stocks	24.1	30.4	17.8	40.7	20.1	17.8
	Growth rate of product export	17.0	20.1	11.8	24.2	14.2	11.8
	Real wage growth rate	8.0	7.5	8.4	8.9	6.0	8.4
	Employment growth rate	3.9	7.6	0.2	9.0	6.1	0.2
	Profit ratio(P/K)	0.20	0.24	0.14	0.29	0.18	0.14
	Profit distribution(P/Y)	0.60	0.62	0.55	0.64	0.59	0.55
	Output/ capital(Y/K)	0.34	0.38	0.25	0.44	0.30	0.25
	Growth rate of labor productivity(Y/L)	7.1	6.8	7.5	7.0	6.6	7.5
	Growth rate of capital intensity(K/L)	10.0	9.2	11.6	10.1	7.8	11.6

Table 1- 1 Major Economic indicators

Source: Bank of Korea

1) All indicators except for unemployment, profitability, output/ capital ratio are rates of increase.

2) Rates of increase are calculated based on annual average values.

3) Real GDP increase, Growth rate of Total fixed capital formation, Growth rate of fixed investment, Growth rate of private consumption, export increase, inflation, manufacturing growth rate, Growth rate of Total fixed capital formation, export increase, and real wage increase are based on invariable 1995 price.

4) Inflation rates are based on 1995 GDP deflator.

5) Profit figures and output/ capital figures are current prices.

6) Labor productivity is based on 1995 per labor value addproduction,

Capital intensity is based on 1995 invariable fixed capital stockper worker.

7) Manufacturing real wage was calculated as nominal wages divided by consumer price index.

8) Employment increase in manufacturing is the monthly average increase rate in manufacturing.

9) Government production of services are excluded.

Second, private sector consumption has been consistently increasing, while export growth rate has been in decline. Since total production from manufacturing would be consumed either in export or domestic markets, relative decline in export markets would, in turn, mean the importance of domestic market has been increased.

Third, since the late 1980s on real manufacturing wages in Korea have in sharp increase.

Fourth, during the late 1980s, profit rates in Korean manufacturing sectors were reduced by 10%. If one breaks down profit elements into Profit distribution(P/Y) and Output/ capital(Y/K), it is clear that the former has been erduced by 7% ,while the latter was educed by 13% during the same period. While both Profit distribution(P/Y) and Output/ capital(Y/K) are in decline, one can notice that Output/ capital(Y/K) is more responsible for the decline in profit figures (Sakong 1987; Kim, J. 2005).

If one breaks down Output/ capital(Y/K) into two sub elements of Growth rate of labor productivity(Y/L) and Growth rate of capital intensity(K/L), it is possible to infer that since the late 1980s on, the increase of capital intensity was far faster than that of labor productivity. This could be the main reason for the decrease in profit figures. This does not necessarily mean labor productivity figures went bad in the time period. Rather, it is fair to argue that while labor productivity figures were decent, upgrading the economy and manufacturing sector itself was so fast that the figures in capital intensity has gone up so high, which eventually resulted in the decline of profit numbers. From a different angle, this is viewed as a façade of a "productivity paradox" phenomenon in a benign context (Kim, J. 2005).

The advent of Knowledge Economy

With the general trend of transition just set up in the preceding section, it is possible to dig into a layer of knowledge economy. As seen in the table below, in terms of contribution to productivity increase in Korea, the relative importance of manufacturing has been weakened from 73.5% during 1995-2000 period to 58.8% during 2000- 2004 period (Kim, J. 2008). Between knowledge intensive service sector and the traditional service, the so-called high flyers had stronger contribution in 2000-2004 period, compared to the comparative number in 1995-2000 period.

Yrs	Contribution to productivity increase (%)				Productivity Increase (%)
	Knowledge Intensive Service	Traditional service	Manufacturing	Others	
1995-2000	25.6	-16.1	73.5	18.1	3.5
2000-2004	31.6	-18.7	58.8	27.1	3.0

Table 1-2 Contribution to productivity increase

Going into details in the tables below, among 3.3 million establishments in 2005, there are about 114,133 knowledge intensive manufacturing firms and around 663,890 knowledge intensive service sector firms in Korea. Among the service firms, which take about one fifth of total establishments, the typology of advanced service sectors was similar to that in other advanced economies, which are mainly composed of finance, logistics, and business services (Hansen 2002).

| | All Sectors | Manu-facturing | Knowledge Service | | | | | |
			Business service	Tele comm	Finance insurance	Distri-Bution (wholesale)	Logistics	Total
# of establish-ments	3,204,809	114,133	84,226	9,028	33,904	207,211	329,521	663,890

Table 1-3 Number of establishments (2005)

Source : Statistical Bureau of Korea

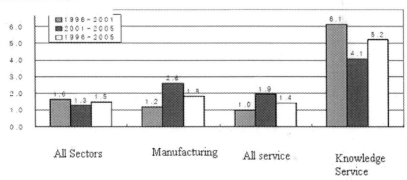

Source: Statistical Bureau of Korea, 2006 wholesale, retail and service sector survey

Figure 1-1 Average increase rate of establishments by major sectors (1996-2005)

As figure shows, in terms of average annual growth, these knowledge intensive sectors took the highest growth rates in all three periods of 1996-2001, 2001-2005, and 1996-2005 periods. Figure shows a trend in selective advanced economies on the relative portion of advanced knowledge services in GDP. Knowing that, usually in most advanced economies, the proportion of service sectors range from 50 to 80 % of GDP, it would be fair to argue that the proportion of knowledge service would be at least half of the total portion of service sectors in GDP in advanced economies.

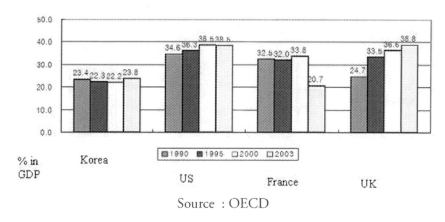

Figure 1-2 Proportion of knowledge based service sector in GDP

If the importance of advanced service sectors can be acknowledged, it is meaningful to see a case example of weather service sector.

Weather / Climate service sector

Weather service sector is a purely information and knowledge based industry, which makes a good example of knowledge intensive sector (McCaffrey 2006). Despite the rosy picture, reality in many knowledge economies, in terms of relative portion of GDP coming from service sectors, shows that this type of industry is in difficult positioning in business point of view. SWOT analysis can help us in explaining why it is the case.

	Strengths • Long relations with Gov't • High level of accessibility • Good understanding of gov't policies	Weaknesses • Difficulty with early market creation • Small scale of business • Potential problem regarding division of labor with gov't
Environments		
Opportunities • Strong growth potentials • Low entry barriers	SO strategy • Maintain good relations with gov't • Respond to gov't policy change • All those wish can enter the market	WO strategy • Should focus on making markets • Limited scale economies should be resoloved.
Threats • Reluctance to pay for weather Info • Weather Market not established	ST strategy • Using access to public, reduce psychological barriers to pay • Try to set up market	WT strategy • Utilize and cooperate with gov't to set up a market

Table 1-4 SWOT analysis of weather marketing

In a typical SWOT analysis, strengths, weakness, opportunities and threat are laid out so that four types of strategies can be presented. For example, SO strategy means a set of strategies that come out from a combination of strengths and opportunities. Exactly identical patterns can be presented for each of the other three strategies.

From the four cells in the table, it is possible to infer that weather service sector suffers from the two major problems. One is the identity of the market. Even in most advanced economies, it is still not clear whether there is a commercial market for weather related services and information. One would argue that looking at media, like TV and newspapers, there is a commercial es, this is true that there is a market in that sense. A real problem in arguing for the existence of a market in this sense comes from how you can actually charge people who used or are willing to use the information.

In the real world, there is a huge discrepancy between what people demand and what people are willing to pay for the weather information. People want to be provided with weather information and services as public goods, while they want to have high quality services.

One would raise the next query by saying that making a market would be a solution to this. While this is a correct approach, there is the second problem the weather service sector has been faced with. It is that there is a psychological barrier in people's minds upon their decision to pay. This means that people would try to hide their true demand in order to pay less. This will eventually bring about under the equilibrium production level of the services and information, since production will react to 'revealed' demands (Olsen 1962).

This is why there have been limited numbers and limited size of weather service sectors across countries. As seen in the table below, the number of private weather service and its related companies are limited in numbers in Japan and the U.S., and especially in the U.S., except for the weather channel, a cabel TV firm whose sales volume exceeds 1 billion U.S. dollars around year 2003-2005 period, remaining numbers are only micro companies which do not have real meanings as absolute numbers (Kim, J.).

Nation	Per capita GDP (US $)	Population (in millions)	GDP (US billions)	# of private weather service firms
Japan	34,375	127	4,349	40
UK	24,549	59	1,442	400 (including small ones)
US	34,047	273	9,299	10 large plus 400 small ones
Korea	8,581	47	500*	9

Table 1-5 Number of private weather service firms and GDP size

Going back to our discussion at the beginning of the chapter, growing of service sectors is an inescapable trend (Chandler 1990). In selected European countries, se was increased service sector growth rates were observed after working hours were cut. Probably, in proportion to the increase of service sectors, it would be fair to think of increased demand for weather service industry (Kim, J. 2007,2008).

**Table 1-6 Annual average growth rate of service
sectors: a before and after comparison**

	Before Working hours reduction	During working hours Reduction		After working hours reduction
Portugal	1992-1996	1996-1997		1997-2001
	-	2.8		4.84
France	1995-1997	1998-2000		2000-2001
	1.70	3.30		2.11
Italy	1994-1997	1997-1998		1998-2001
	1.95	1.96		2.49
Spain	1991-1994	1994-1995		1995-1999
	0.79	2.58		2.93
Japan	1983-1988	1988-1994	1994-1999	1999-2001
	3.92	4.48	3.47	

note : annual average, 1995 invariable price

Source : OECD, Quarterly National Accounts
Bank of Korea , 2001.

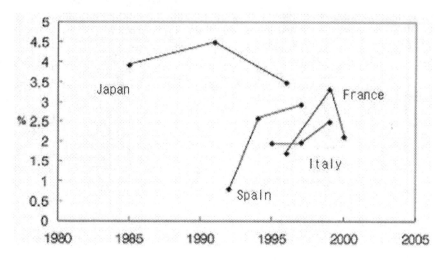

Figure 1-3 GDP growth rate after working hour reduction

Current Division of Labor between gov't and the private sector

Before we move on our discussion on weather service sector, it would be useful to look through how division of labor has been in practice in major countries, as one sees in the table.

		UK	US	Japan	Austrailia
Policy		Private firms exist, but they are thoroughly under MET guidance. -It is fair to understand as the gov't leading model.	-No active gov't Policy for promoting private firms. -Free Service of meteo info is gov't Policy.	-Private firms are Centering around a Non-profit Organization Specialized In meteorology.	Commercial Division(SSU) and private firms compete -Future plans to transfer SSU to private sector when private market becomes visible.
Division Of Labor	Govt	Providing Info with High level Of processing	Forecasting/ Warning info Provides info With basic processing	National Forecast, basic services Are regarded as Gov't domain	Gov't provides basic meteo info to people for free. Value added meteo services Provided by SSU
	Private	Small firms: Reprocessing Basic info from gov't meteo bureau Large firms: Retain inhouse R&D capability	Reprocessing Basic info from gov't meteo bureau	- point forecast Reprocessing Basic info from gov't meteo bureau	Reprocessing Basic info from gov't meteo bureau Extra service Possible with contracts with customers
Charge policy		Commercialization	Free distribution	Partial Commerciali- zation	Partial commerciali- zation
Types pf Private-Public Partnership (PPP)		Public Sector initiating PPP	No Specific PPP	Private sector Initiating PPP	No Specific PPP

Table 1-7 Government Policy toward private weather firms

Phenoenon	Sectors with positive effects	Sectors with negative effects
Hail	Energy	Construction, agriculture
Scouraging weather	Beverage, water	Insurance, energy, agriculture
Increase of sunny days	Leisure, tourism	
Rainfall	Agriculture, hydroeelctric power generation	Beverage, water
Heavy rain	Construction	Agriculture, insurance
Wind	Windfarm	Insurance, airline
Snow	Winter sports	Airline, airport, travel
Tide	Tidal power generation	Fishery, marine transportation, insurance
Fog	-	Airline, logistics

source: Allen Consulting group, *Climate change risk and vulnerability: promoting an efficient adaptation response in Australia*, 2005.3.

Table 1-8 Cost and Benefits of Climate Change

Scope of the Book

After introduction in chapter one, this book will present an empirical survey of industries on weather service and climate change responses in chapter two. Chapter three will have a common theme of 'creating market', in which the chapter will be devoted to an analysis based on BCG matrix of industries to show industry responses to climate change and examples of weather service industries.

Chapter four will investigate policy measures to promote weather and climate related sectors with SWOT analysis, while chapter five will wrap the discussion of the book with possible concluding remarks.

CHAPTER 2

AN EMPIRICAL ANALYSIS
OF INDUSTRY RESPONSE
TO CLIMATE CHANGE

This chapter is based on two empirical surveys of firms in Korea. The first one was conducted in 2008 on their responses to climate change and a separate survey of about 300 firms was exercised in 2007 for both manufacturing and service sectors on their willingness to pay for meteorological services or long term contract research for their business.

The 2008 survey is unique in the sense that it was a CEO survey to find out firms' intention in strategic vision. In comparison, the 2007 survey was a comparatively large company sample survey of 300 firms in both manufacturing and service sectors.

1. Responses to Climate Change

In the 2008 survey, 59 CEOs responded in a survey questionnaire during July. As the table shows, among the 59 firms, 36 were manufacturing, while the remaining 22 firms were from service sectors. Of the three firms that did not identify themselves as either manufacturing or service sectors were basically service firms closely linked to manufacturing process. This could be regarded as a way of adapting to new environments.

	Freq	%	% without non-response	Cumulative %
Metal	6	10.2	10.7	10.7
Machinery	5	8.5	8.9	19.6
Electronics	3	5.1	5.4	25.0
Computer	2	3.4	3.6	28.6
Software	3	5.1	5.4	33.9
Construction	6	10.2	10.7	44.6
Chemical	4	6.8	7.1	51.8
Textile/ apparel	4	6.8	7.1	58.9
Business service	15	25.4	26.8	85.7
Service (sales)	7	11.9	12.5	98.2
Others	1	1.7	1.8	100.0
Sub-Total	56	94.9	100.0	
No response	3	5.1		
Total	59	100.0		

Table 2-1 Industries by type

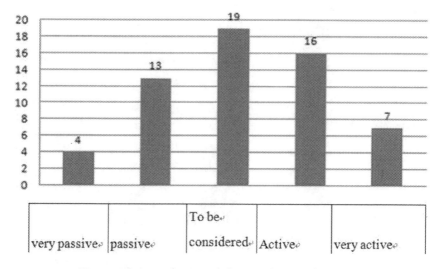

Figure 2-1 Attitude toward climate change adaptation

Preparation for Climate Change

When the CEOs were asked whether their firms were exercising climate change adaptation efforts, about 61% was shown as passive in a borader sense, while 39% of firms were shown as active in preparation for climate change adaptation.

ii	freq	%	% without no-response	Cumulative %
Very passive	4	6.8	6.8	6.8
Passive	13	22.0	22.0	28.8
To be considered	19	32.2	32.2	61.0
Active	16	27.1	27.1	88.1
Very active	7	11.9	11.9	100.0
Total	59	100.0	100.0	ii

Table 2-2 Efforts toward climate change adaptation

Forecasting the Time Frame

In a question "when do you think climate change adaptation would be an important task, 56.9% of firms expressed that within 3 years climate change adaptation would become an implementation task. If one adds up to 5 years, it would be 84.5% of CEOs agreeing on the imminence of the issue.

Time frame	Freq	%	Cum %
1 yr from now	11	18.6	19.0
3 yrs from now	22	37.3	56.9
5 yrs from now	16	27.1	84.5
7 yrs from now	5	8.5	93.1
10 yrs from now	4	6.8	100.0
Sub total	58	98.3	
No response	1	1.7	
Total	59	100.0	

**Table 2-3 when do you think climate change
adaptation would be an important task?**

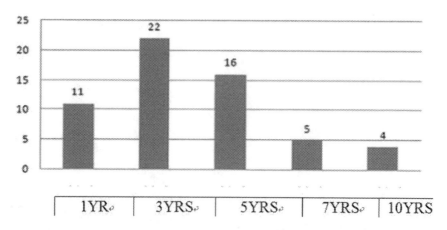

Figure 2-2 Time frame when climate change adaptation will be important

	FREQUENCY	%	% with effective responses	Cumulative %
0%	5	8.5	8.6	8.6
25%	26	44.1	44.8	53.4
50%	14	23.7	24.1	77.6
75%	11	18.6	19.0	96.6
100%	2	3.4	3.4	100.0
SUBTOTAL	58	98.3	100.0	
NO RESPONSE	1	1.7		
TOTAL	59	100.0		

**Table 2-4 Probability that climate change adaptation
will be an important competitiveness factor**

In a slightly different format, when CEOs were asked "what would be your subjective probability that climate change adaptation would be a key competitiveness factor for your firm and the industry where your firm in located?, 46.5% of CEOs, who answered one of the three answers of 100%, 75%, or 50%, showed that they were thinking it as an important competitiveness factor.

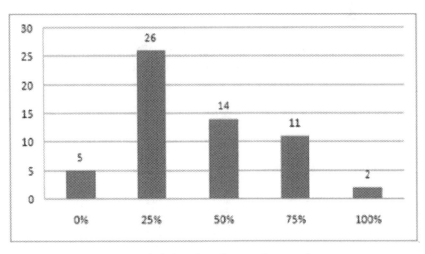

**Figure 2-3 Probability that climate change adaptation
will be an important competitiveness factor**

In a differentiated question with exact time frame, the survey asked the CEOs whether they consider climate change adaptation as a key competitiveness factor in 10 years. This question is different from the preceding one in that it gave a time frame. On this question, 64.4% of respondents said it would be important by counting the CEOs who answered for 50, 75, and 100%. By giving 10 year time frame, CEOs gave higher percentage in their perception for climate change as an important strategic factor. It was interesting to find that about 34.5% of firms either expressed as 0% or 25% for climate change becoming the key competitiveness factor. In figure , answers for the 10 year time frame featured a typical bell shape normal distribution like symmetry in CEOs answers.

	Freq	%	% with effective responses	Cumulative %
0%	5	8.5	8.6	8.6
25%	15	25.4	25.9	34.5
50%	19	32.2	32.8	67.2
75%	15	25.4	25.9	93.1
100%	4	6.8	6.9	100.0
Subtotal	58	98.3	100.0	
No response	1	1.7		
Total	59	100.0		

Table 2-5 Probability that climate change adaptation would be an important competitiveness factor in 10 year

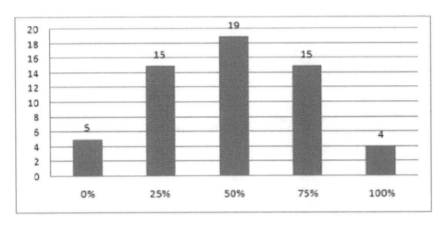

Figure 2-4 Probability that climate change adaptation would be an important competitiveness factor in 10 year

Are there your competing overseas firms?

On this question, except for the firms that did not answer, about 76% of firms in the survey had their competing firms in the international markets.

	FREQ	%	% with effective responses	Cumulative %
YES	39	66.1	76.5	76.5
NO	12	20.3	23.5	100.0
SUBTOTAL	51	86.4	100.0	
NO RESPONSE	8	13.6		
TOTAL	59	100.0		

Table 2-6 Overseas competing firms

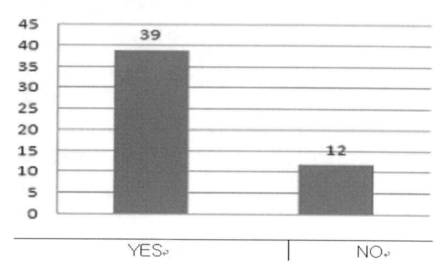

Figure 2-5 Overseas competing firms

Gap between your firm and competing overseas firms

CEOs were asked on they perceive the gap between theirs and their competing firms in terms of their preparation for climate change adaptation. As seen in table in the below, there was an even distribution of answers. Remarkably about 21% of firms expressed that there was no gap, indicating that the firms in the survey were quite advanced ones. Answers for 3 years, 5 years, and 10 years all showed percentages in 20s.

	freq	%	% with responses	Cumulative %
No gap	8	13.6	21.6	21.6
1 yr	3	5.1	8.1	29.7
3 yrs	9	15.3	24.3	54.1
5 yrs	9	15.3	24.3	78.4
10 yrs	8	13.6	21.6	100.0
Sub total	37	62.7	100.0	
No response	22	37.3		
Total	59	100.0		

Table 2-7 Time gap with foreign firms

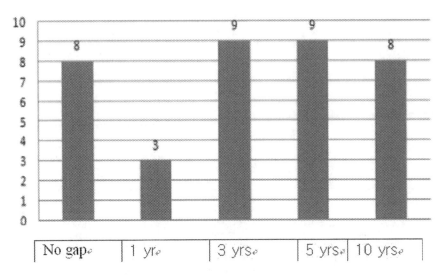

Figure 2-6 Time gap with foreign firms

Technology development fields for Climate Change Adaptation

CEOs were asked a broad question regarding what kinds of technology items should be developed or introduced to Korean markets of your fields. They were asked to give specific years remaining until they are to be introduced either by developing or importing. Table below shows a list of technology items both in manufacturing and service sectors.

Types of business	Items for development	Period for development or import
Metal	Automobile weight reduction	1 yr after
	Lubricant oil for cutting	3 yrs after
	Electric furnace replacement	10 yrs after
	High efficiency heat pipe	immediately
	Heat recycling system	1 yr after
	Energy saving heater	immediately
Computer	Heat diffusion for motors	5 yrs after
Electronics	Clean room facilities	10 yrs after
	Vehicle emission eduction technology	5 yrs after
	Metal gas processing and enhancement	2 yrs after
	Flexible display equipment	3 yrs after
Software	Energy saving through software	immediately
Construction	Improving solar collector	1 yr after
	Heating / cooling system improvement	5 yrs after
	Environment friendly construction material	5 yrs after
	Outdoor work environment improvement	3 yrs after
Machinery	Food container	5 yrs after
	Hybrid vehicle	1 yr after
	Fuel cell powered vehicle	10 yrs after
	Transmission for hybrid vehicles	1 yr after
	Transmission for fuel cell vehicles	6 yrs after
	Electric car	immediately
	Electric bike	1 yr after
	3 wheeled electric bike	1 yr after
	14 passenger solar hybrid vehicle	5 yrs after

Chemical	Printer recycling technology	1 yr after
	Energy recycling technology	3 yrs after
	Parts development for fuel cell engine system	5 yrs after
Textile/ apparel	Ultra light high performance textile	1 yr after
	Winter shoes	5 yrs after
	UV protection clothes and swimwear	1 yr after
	Body temperature control apparel	5 yrs after
	100% recyclable paper based clothes	5 yrs after
Business service	Environment, health consulting	1 yr after
	CO2 reduction consulting	5 yrs after
	Heat efficiency improvement fo boilers	3 yrs after
	Recyclable energy consulting	1 yr after
	Waste treatment business	immediately
	ESCO	immediately
	Financial derivatives for agricultural products	immediately
	Financial derivatives for raw material	immediately
	Using Energy efficient commercial aircraft	immediately
	Reducing weights in aircrafts	immediately
	Energy saving charters	immediately
	Termal power generation consulting	1 yr after
	World wide weather forecasting system	3 yrs after
	Climate based financial products	1 yr after
	Consulting for physical conditions based on weather	1 yr after
	Schedule management for travelers	1 yr after
Coal	coal utilization technology	15 yrs after

Table 2-8 Items for development

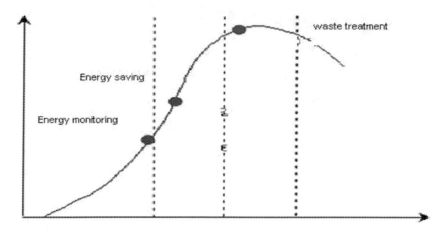

Figure 2-7 An example of climate change adaptation in energy sector

2. Willingness to pay for Weather Services

In the second survey to be introduced in this chapter was a firm survey conducted in 2007 to find out the types of meteorological information firms need and their willingness to pay for the service if the services go private. The sample size of the survey was 308 firms in 23 manufacturing and services sectors, and the firms included in the survey were major firms in Korea in order to secure validity of the research. In the table below, it is possible to distinguish three types of industrial sectors that were surveyed. In each of the questions, the survey undertook a hypothesis testing to see whether manufacturing and service sectors would be different statistically.[1]

	Manufacturing	Traditional service	Knowledge service
Industry	machinery household supplies apparel food	construction material transportation logistics/ delivery transportation service lodging recreation public utilities	Financial services education energy environmental service tourism business services consulting software development

Table 2-9 Sectors in survey

---

1   In this section, among diverse survey questions, selected ones will be presented.

The most important meteorological info for firms

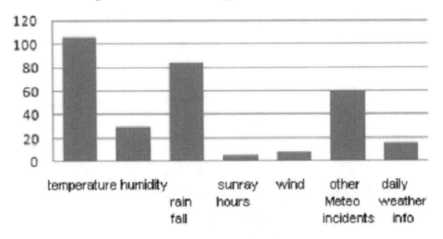

Figure 2-8 The most important meteorological info for firms

> Null hypothesis : There are no difference between manufacturing and service sector response patterns.
>
> Alternative Hypothesis : There are difference between manufacturing and service sector response patterns.

On this question, 106 firms, 42 manufacturing and 64 service sector firms responded that temperature is the most important weather information. The second highest response was rainfall, followed by other meteorological phenomenon like thunder, lightening, and special weather events like hurricane and tornado. Implications from this were very clear. Not different from conventional thinking, temperature and rain were the most sought after information for most firms. In comparison, whenever there are special meteorological events, the value of weather information goes up exponentially, and this is reflected in the answer pattern.

Responses	temper-ature	humidity	rainfall	sun ray hours	wind	other meteo incidents	daily weather info	total
Manufac-turing	42	23	11	1	0	11	5	93
Service	64	7	73	4	8	49	10	215
Total	106	30	84	5	8	60	15	308

Table 2-10 The most important meteorological info for firms

In terms of statistical difference between manufacturing and service sectors, it turned out that there is a difference between the two.

How often do you retrieve weather info?

When firms were asked how often they retrieve weather information, answers for 'once a day' marked the outstanding answer type, followed far behind by 'whenever needed' excluding no answers.

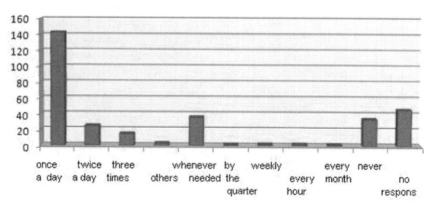

Figure 2-9 How often do you retrieve weather info?

Response	once a day	twice a day	3 times a day	others	whenever necessary	3 times a quarter	once a week	every hour	every month		no response	total
Manufacturing	50	7	1	2	8	1	2	0	1	8	13	93
Service	93	18	14	1	28	0	0	2	0	26	33	215
Total	143	25	15	3	36	1	2	2	1	34	46	308

Table 2-11 How often do you retrieve weather info?

Statistical testing showed that manufacturing and service sectors showed statistically meaningful difference in Chi-square testing.

Is long term forecast useful for your industry?

One of the curious question in designing survey was whether firms need and regard 'long term forecast' as important element in planning their business. Also interesting was how long would be understood as the 'long term' in each of the industry.

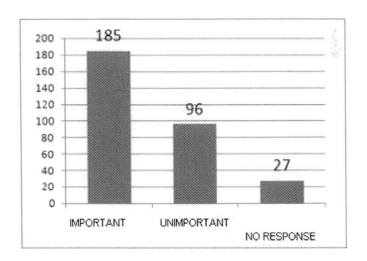

Figure 2-10 Is long term forecast useful for your industry?

About 60% of firms responded that long term forecast is important, while remaining 31.2% answered as unimportant and 27 firms avoided answering. When manufacturing and service sectors were tested statistically, the two showed statistically significant difference. Possible interpretation from the frequency distribution was that while similar percentage of firms in each of manufacturing and service sectors answered long term forecast as important, percentage of non response firms were higher in manufacturing sector firms.

Response	important	unimportant	no response	total
Manufacturing	49	18	26	93
Service	136	78	1	215
Total	185	96	27	308

Table 2-12 Is long term forecast useful for your industry?

How far is your long term forecast?

If firms have responded that long term forecast is an important element, then a following question would be that 'how long would be considered as 'long term'. One month was the highest ranking answer, followed by 3 months. Looking into cell of frequency, for '1 month' answer choice, service firms marked more in sheer numbers. Considering total number of manufacturing firms answered compared to the total size of service firms, even from a 'naked eyes', the two sectors show similar pattern. Statistical testing showed that the two sectors had no difference.

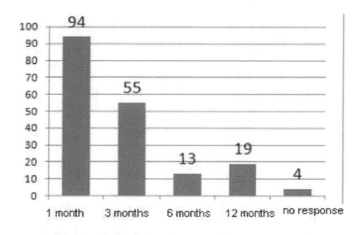

Figure 2-11 How far is your long term forecast?

30

Response	1 month	3 months	6 months	12 months	no response	total
Manufacturing	23	17	4	3	2	49
Service	71	38	9	16	2	136
Total	94	55	13	19	4	185

Table 2-13 How far is your long term forecast?

Which channel do you use most for weather information?

This question asked which channel would be industry's favored choice for receiving weather information. This question was intended for a 'litmus test' for a future change. On this question, a remarkable pattern was observed. Unlike a conventional expectation that firms would prefer government channel, the meteorological administration, firms expressed that they prefer getting weather information from web portals and other new media.

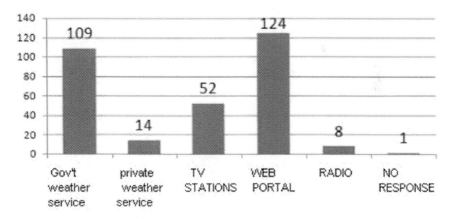

Figure 2-12 Favored channel

Statistical testing showed that the two sector of service and manufacturing did not show difference.

Response						no response	total
Manufacturing	36	7	16	29	4	1	93
Service	73	7	36	95	4	0	215
Total	109	14	52	124	8	1	308

Table 2-14 Favored channel

In planning your business decisions how much confidence do you have in weather forecast?

It would be also an interesting question to know how much confidence would firms have in using weather information. 41.9% of firms answered as less than 50%, followed by 36.7% with between 50-70%. Both manufacturing and service sectors showed no statistical difference, although service firms were concentrated more in the 'less than 50%' choice, while manufacturing firms were almost equally distributed in 'less than 50%' and between '50-70%' choices.

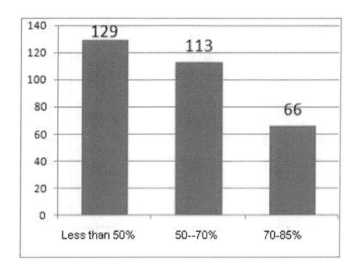

Figure 2-13 Confidence in weather forecast

Response	50% or less	50%-70%	70-85%	Total
Manufacturing	38	40	15	93
Service	91	73	51	215
Total	129	113	66	308

Table 2-15 Confidence in weather forecast

How much contribution do you expect from weather info to your revenue?

This question would be a quintessential one in understanding the value of meteorology in real world business. The question was directly asking how much percentage could be contributed to weather portion in each firm's revenue generation. 33.8% of firms responded as between 1-5%, followed by 19.2% of firms(59 firms) for between 6-10%. It is noteworthy that among firms that answered as others, 21 firms out of total of 308 firms responded as less than 1%. 13 companies among 'others' reported as 50%, while 2 firms were considered as outliers in the sense that they answered as 100%.

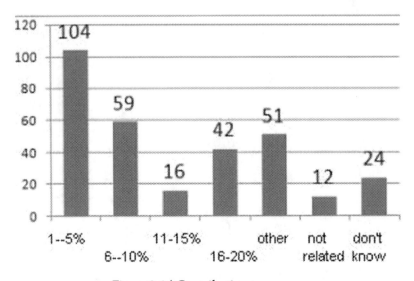

Figure 2-14 Contribution to revenue

Response	1-5%	6-10%	11-15%	16-20%	others	Not related	no response	total
Manu-facturing	37	23	4	11	0	0	6	93
Service	67	36	12	31	2	12	18	215
Total	104	59	16	42	2	12	24	308

Table 2-16 Contribution to revenue

It would be sensible that firms were considering weather in not that directly linked to revenue generation. Statistical testing showed manufacturing and service sectors showed no difference.

Time gap between the acquisition of long term info and actual usage

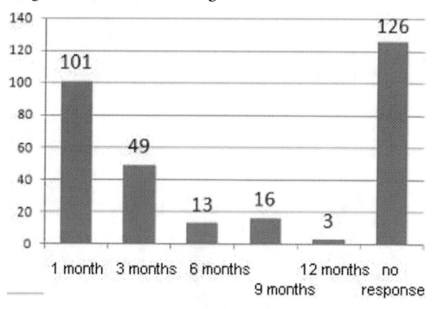

Figure 2-15 Time gap until usage

A question was given on the time gap between when firms get long term weather information and the actual time frame thay would use that information.

Response	1 month	3 months	6 months	12 months	others	no response	total
Manu-facturing	25	14	4	1	0	49	93
Service	76	35	9	15	3	63	215
Total	101	49	13	16	3	112	308

Table 2-17 Time gap until usage

'No response' was the highest answer, followed by '1 month'. This shows congruence with response patterns from a question 'how long would be your long term weather forecast?' in previous question list. Firms with no response could be interpreted as having unclear definition of long term weather forecast in business or having difficulty in applying meteorological information. Both service and manufacturing sectors showed no statistical difference in terms of response patterns.

What would be your dissatisfaction with weather service?

Since some of the reasons for not using weather information may come from industries' dissatisfaction on weather service itself, a question was given to address this issue. This hit the 'target' in the sense that a overwhelmingly high percentage of answers were concentrated in 'low accuracy', while all other answer choices showed similar frequency.

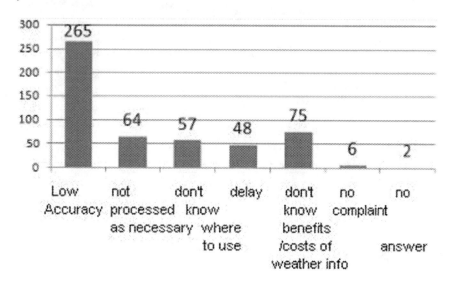

Figure 2-16 dissatisfaction with weather service

While the most important 'confirmation' was the low accuracy, other answers also give very salient clues for the usage of weather information. Other choices, like 'not processed as necessary', 'don't know where to use', and 'don't know benefits and costs of weather info', all are interwounded. In some sense, because weather information is not processed easily enough for firms to use, they are not using. Because they do not know where to use and costs/benefits of weather exactly, they don't make efforts to process information as they needed. These reasons are self-reinforcing within themselves, which turns out to be a huge hiatus between providers' side and consumers' side.

Response	Low accuracy	can not be processed as needed	do not know where to use	delays in delivery	no knowledge on benefits /costs of meteorological information	no response	total
Manu-facturing	73	5	9	1	1	2	2	93
Service	192	7	5	3	4	4	0	215
Total	265	12	14	4	5	6	2	308

Table 2-18 Dissatisfaction with weather service

Statistical testing for this question showed that service and manufacturing sectors revealed statistical difference. It turned out that more service firms are sensitive to inaccurate weather information. This reminds one of the previous response pattern on the dissatisfaction on weather service. While service firms do recognize that they are sensitive against weather information, they are not fully aware of how to utilize, and process weather information as they needed. Furthermore, they may be in trouble in understanding who will be paying for the additional costs for 'processing.

Do you know how weather info is utilized in firms of the same industry in other countries?

This question was intended to check whether firms have reference points in competing firms in other countries. It turned out that most firms did not have information on this. Statistical testing showed statistically meaningful difference between manufacturing and service sector firms. Service firms were more aware of their foreign competitors's cases than their manufacturing counterparts.

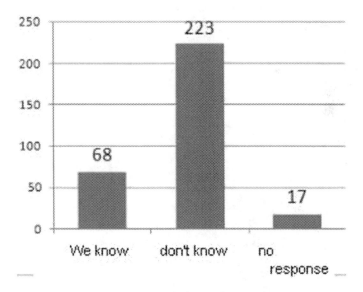

Figure 2-17 Knowledge on foreign firms

Response	knew	don't know	no response	total
Manufacturing	17	59	17	93
Service	51	164	0	215
Total	68	223	17	308

Table 2-19 Knowledge on foreign firms

Intention to invest in using meteorological information at your firm's activity

This would be one of the quintessential questions in evaluating whether weather market can be created. Among samples, 56.5% of firms responded negatively, while there was statistical difference between manufacturing and service. In interpreting this, however, it is possible to see the opposite side in the sense that the other 40 plus percentage of firms showed positive responses. This may be a starting point from which marketing can be initiated.

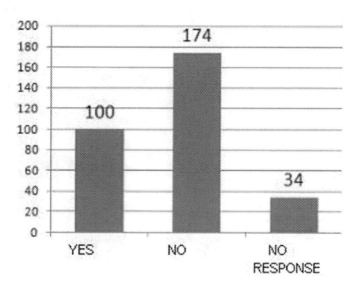

Figure 2-18 Intention to invest

Response	YES	NO	no response	total
Manufacturing	26	46	21	93
Service	74	128	13	215
Total	100	174	34	308

Table 2-20 Intention to invest

Intention to use R&D budget in investing in utilizing weather information

As a follow up question, a question was given to ask firms' intention to utilize their R&D resources into meteorological application for their business. On this question, almost half split answers were gleaned with no statistical difference between the two sectors of manufacturing and service.

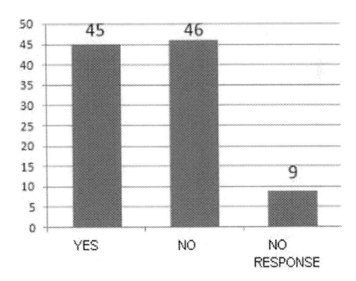

Figure 2-19 Intention to use R&D budget

Response	Yes	No	no response	total
Manufacturing	12	11	3	26
Service	33	35	6	74
Total	45	46	9	100

Table 2-21 Intention to use R&D budget

Proportion of investment

For the firms that expressed their intention to invest in weather application for their business, a question was given on the proportion of investment they can allocate. As one sees, between 0.1-1% took the biggest share, followed by no response. 1 to 2% took the second largest answer choice excluding no response.

Manufacturing and service sectors did not show statistical difference in terms of distribution of answers.

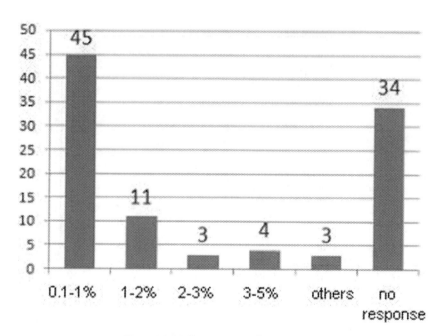

Figure 2-20 Proportion of investment

Response	0.1-1%	1-2%	2-3%	3-5%	others	no response	total
Manu-facturing	12	5	0	0	1	8	26
Service	34	6	3	4	0	26	74
Total	46	11	3	4	1	34	100

Table 2-22 Proportion of investment

9 Major Industries

In this section, among 23 industrial sectors surveyed, 9 sectors that had enough samples were focused for a further analysis at individual industry level.

Everyday household products

This industry represents a variety of sectors that produce everyday household products, which can be renamed as light industries in traditional taxanomy. In this industrial sector, the most important meteorological phenomenon was rainfall, followed by temperature. This can be easily understood in the sense that the nature of products from this industrial sector is heavily related to activities of people under weather.

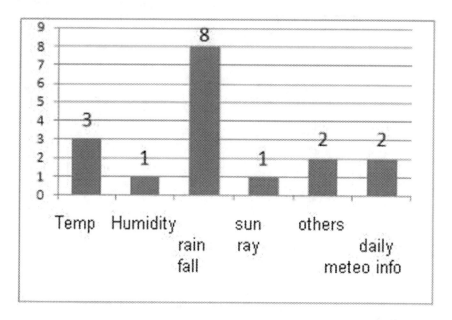

Figure 2-21 Important weather phenomenon: *Everyday household products*

On the question which distribution is favored most in receiving weather information, for this sector, government, which is Korea meteorology administration, was scored the highest.

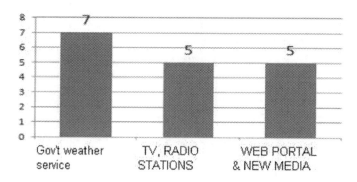

Figure 2-22 Favored channel: *Everyday household products*

When asked about the reasons for possible dissatisfaction on weather information, firms in this sector replied that low accuracy as the prime factor.

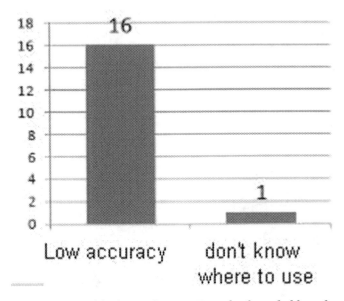

Figure 2-23 Possible dissatisfaction: *Everyday household products*

This sector responded that 35% of its firms under the survey were willing to invest in weather related application for their business.

Food & Beverage Industry

This sector is also one of light industrial sectors in conventional classification of industries. For this sector, temperature turned out to be the most significant weather factor for their business, followed by daily weather information and special meteorological events.

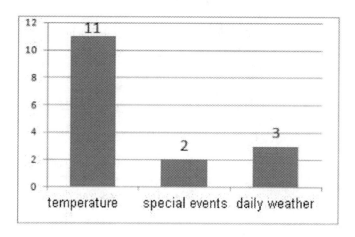

Figure 2-24 Important weather phenomenon: *Food & Beverage Industry*

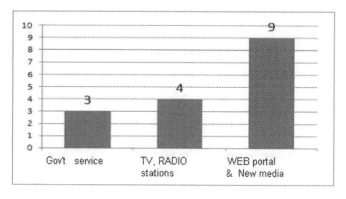

Figure 2-25 Favored channel: *Food & Beverage Industry*

Regarding their most preferred distribution channel, the sector answered that web portal and the new media as their preferred path, while government channel was less preferred. This could be interpreted as a nature that comes from the characteristics of the sector, which are closely related to consumption pattern of people being closely related to new media (Weiss 2007; Wittwer 1995).

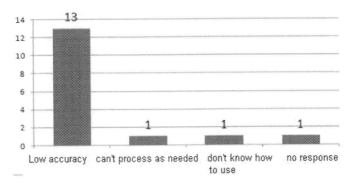

Figure 2-26 Possible dissatisfaction: *Food & Beverage Industry*

As for their dissatisfaction factors, the sector responded that low accuracy being the first factor (Burt 2007), while others were similar in their frequency.

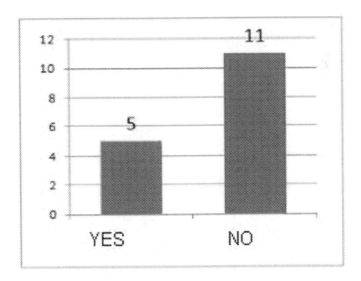

Figure 2-27 Intention to invest: *Food & Beverage Industry*

Food and beverage sector expressed its relecutance to invest in weather information.

Construction

Construction sector showed that rainfall was the top most interested meteorological phenomenon, followed by temperature. This may be due from thenature of the work this industry is engaged in. If it rains or the weather is humid, construction sector would experience delays and this would mean increased costs.

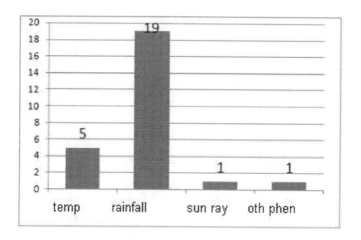

Figure 2-28 Important weather phenomenon: *Construction*

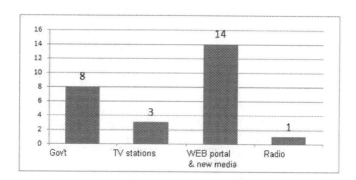

Figure 2-29 Favored channel: *Construction*

Construction sector showed preference on new media like web portals, followed by traditional government channels. This is a repeated pattern in different sectors.

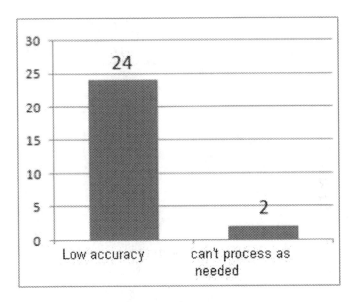

Figure 2-30 Possible dissatisfaction: *Construction*

Like other sectors, construction sector also showed that low accuracy was the source for dissatisfaction. Reflecting industrial needs, the sector expressed that weather information was not processed as they needed. On willingness to invest in R&D, 69% of firms declined to invest in weather application. This is against traditional nature of the sector as well as against practical necessity. In interpretation, it would be reasonable to infer that the sector has strong incentives to hide its preference to be benefited from weather information (Underwood 2004).

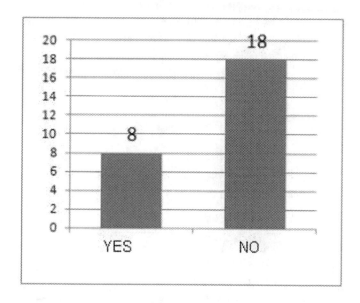

Figure 2-31 Intention to invest: *Construction*

Raw Material

This sector was found to be sensitive against temperature and rain, together with other special phenomenon like hurricane, tonado, and lightening.

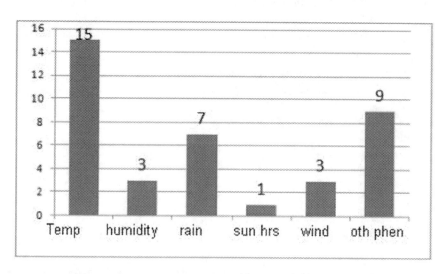

Figure 2-32 Important weather phenomenon: *Raw Material*

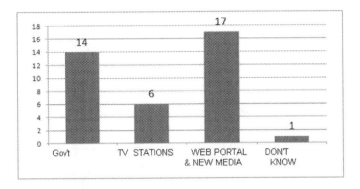

Figure 2-33 Favored channel: *Raw Material*

Similar to other sectors, this sector also showed its preference toward web portal and new media for its delivery channels. Also identical was low accuracy of weather information. Against this, about 50% expressed that they were not interested in investing in weather related investment.

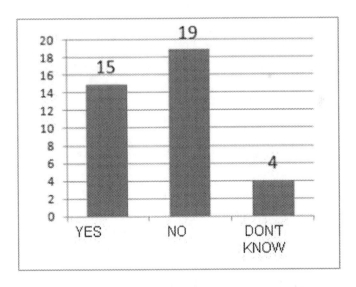

Figure 2-34 Intention to invest: *Raw Material*

Transportation

This sector showed its unique characteristics in its sensitivity toward weather information. Unlike most sectors that expressed temperature or rain as their primary concerns, this sector showed its strong sensitivity toward other specia l weather phenomenon.

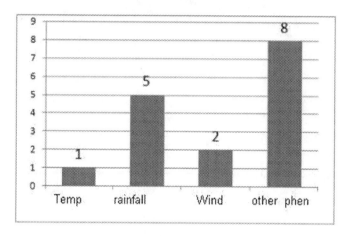

Figure 2-35 Important weather phenomenon: *Transportation*

Also in terms of their preference on weather information distribution channels. this sector showed its preference on government channel, contrast to other sectors' preferred choice for new media. The second preferred channel was remarkably private weather service providers, while new media marked low.

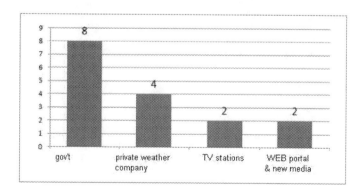

Figure 2-36 Favored channel: *Transportation*

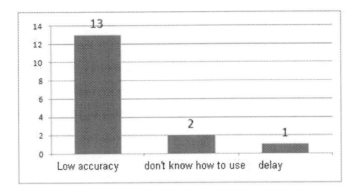

Figure 2-37 Possible dissatisfaction: *Transportation*

For transportation sector, low accuracy was the biggest dissatisfaction element, while intention to invest in weather application for business showed an evenly split answers.

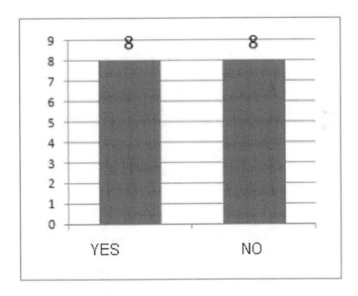

Figure 2-38 Intention to invest: *Transportation*

Hotel & restaurant

Obviously, hotel and restraurant sectors were heavily affected by rainfall, followed almost evenly by temperature, wind, humidity, daily weather information and special weather events. The sector was traditional in the sense that the sector preferred government channel over new media like web portals as were the cases in other sectors.

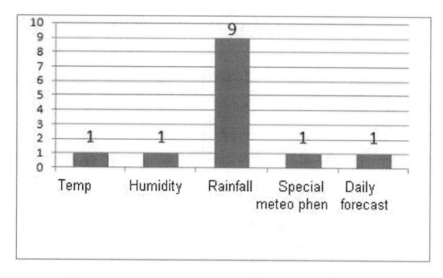

Figure 2-39 Important weather phenomenon: *Hotel & restaurant*

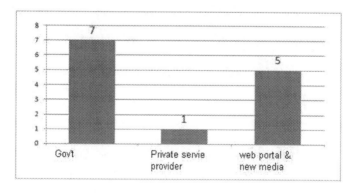

Figure 2-40 Favored Channel: *Hotel & restaurant*

What was common in the sector vis-à-vis other sectors was the 'low accuracy' reason for dissatisfaction. Hotel sector showed reluctance to invest as shown in table below. For the industry, it would be rational for them to be passive in accepting weather information.

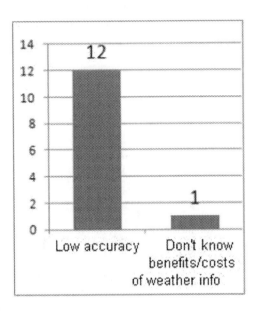

Figure 2-41 Possible Dissatisfaction: *Hotel & restaurant*

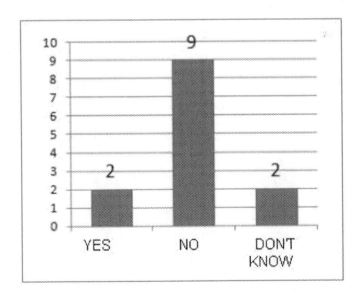

Figure 2-42 Intention to invest: *Hotel & restaurant*

Distribution Service/ logistics

Distribution and logistics sector responded that temperature and rainfall were the top two important weather information. Like most other sectors, the sector showed its preference on web portal and other new media channels over traditional government meteorology bureau. But the highest marked distribution channel was T and Radio stations.

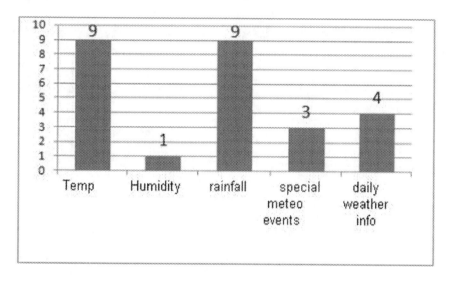

Figure 2-43 Important weather phenomenon: *Distribution Service/ logistics*

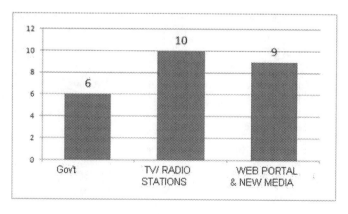

Figure 2-44 Favored Channel: *Distribution Service/ logistics*

When this sector was asked of the reasons of dissatisfaction, the sector listed low accuracy as the prime cause, which has been the single most picked reason. On the intention to invest, about 72% of firms answered negatively.

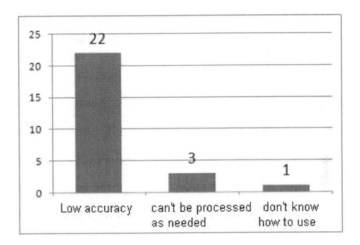

Figure 2-45 Possible Dissatisfaction: *Distribution Service/ logistics*

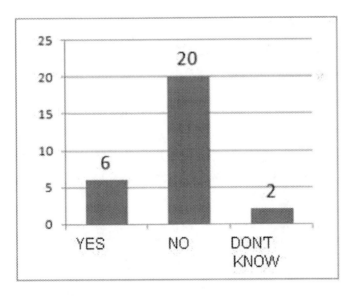

Figure 2-46 Intention to invest: *Distribution Service/ logistics*

Public enterprises

In this survey, large public enterprises were included. They answered that temperature and rainfall were the two most important phenomenon.

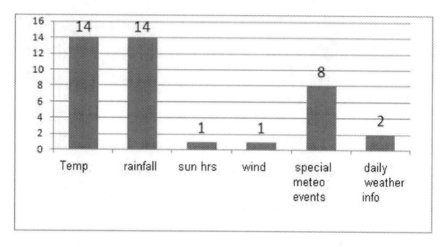

Figure 2-47 Important weather phenomenon: *Public enterprises*

Public enterprises responded that web portal and new media are the most preferred channels for receiving weather service information, followed by government channels. This is quite similar to other sectors in discussion in the previous sections.

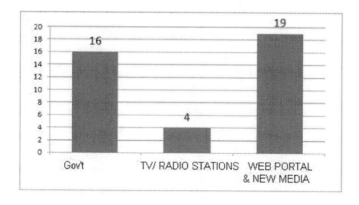

Figure 2-48 Favored Channel: *Public enterprises*

In terms of dissatisfaction, public enterprises were no different from other sectors in the sense that they also reported low accuracy as the prime reason for dissatisfaction. When asked about their willingness to invest in weather information for their business areas, 58% answered negatively.

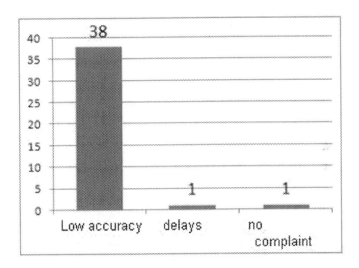

Figure 2-49 Possible Dissatisfaction: *Public enterprises*

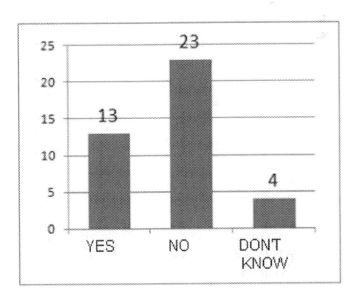

Figure 2-50 Intention to Invest: *Public enterprises*

Finance

Finance industries may be called as the crown jewel sector that may be benefited from weather information. While this may be the dream, reality in survey result was a cold reflection of today. As one can grasp easily, the sector has its deep hands plunged in so called future's market and other derivatives in advanced economies. Even not so active like in the U.S., according to survey results in Korea, still this sector showed some resemblance to the nature of finance sector in advanced nations.

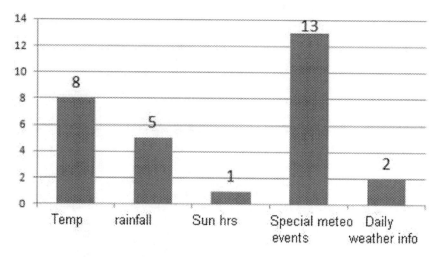

Figure 2-51 Important weather phenomenon: *Finance*

For the finance sector, the most concerning weather information was special meteorological events like tonados, typoons, hurricanes, and others. This is clearly understandable in the sense that the special events are the ones that can change their profit structure completely.

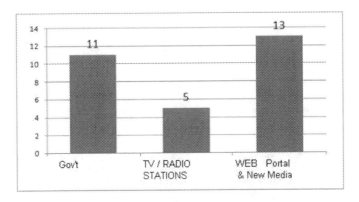

Figure 2-52 Favored Channel: *Finance*

For preference, the sector showed similar preferernce in the sense that web portal and new media surpassed their preference votes for government channels. Also low accuracy issues has been the culprit for dissatisfaction. 52% of finance sector firms responded that they have no plans to invest in weather areas.

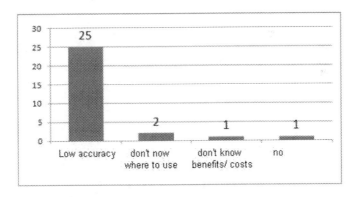

Figure 2-53 Possible Dissatisfaction: *Finance*

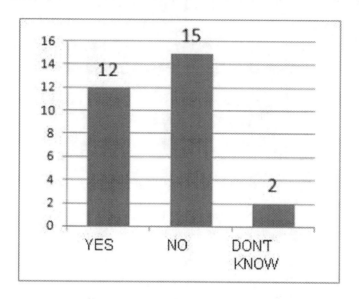

Figure 2-54 Intention to Invest: *Finance*

CHAPTER 3
MULTINATIONAL CASES
OF CLIMATE CHANGE
ADAPTATION AND
WEATHER UTILIZATION

1. Climate adaptation & Weather utilization cases from Korean industries

1) Weather utilization cases from Korea industries

Textile & Apparel

In this industrial sector, an in-depth interview was held at the Shin-Young Wacoru Inc., a famous brand maker for women's underwear. Reflecting the nature of clothes, sales of underwear increases with the lowering of temperature, while in production side, preparing special fabric for summer would be an example of weather utilization.

When sales side is compared to production side, sales is more sensitive against weather. Knowing this, even though the firm tries to sell specially processed products that can be suitable for sultry weather in summer, winter is always a better season to sell products. Except for temperature, other meteorological factors, like heavy rain, storm, and typoon, are relatively unaffecting the sector, since other factors commonly have one single effect upon consumers of products, that is not going out when weather is bad. Thus, temperature is important to this industry, and between hot and cold weather, cold weather is more critical.

One thing to be reminded of in thinking about the impact of weather on apparel industry is that weather may not be a critical factor vis-à-vis the factor of fashion or trend. According to the interview, this point would be amplified in clothes except underwear. For example, in one winter, a specific fabric was great on demand, not because the fabric can protect people from cold well, but because of mere trend in that season. In this sense, short term weather would not affect people's decision to buy clothes. You would not buy clothes to be prepared for an incoming storm. A critical implication would be that this industry would be affected by long term weather or climate. In a more detailed sense, climate change would be the affecting factor.

Food & Beverage

In this survey, an interview was implemented at a food & beverage manufacturer, Lotte-Samgang, which is famous for icecream manufacturing. As one sees in figure 3-1, there is an industry wide known relationship between temperature and the sales volume of ice cream. Until the temperature reaches 30 celcius or 90 F, sales continue to grow. After that there is a saturation toward a certain stage. Industry reports that above 30 celcius, it would be beverage products that would be sold more and not ice cream.

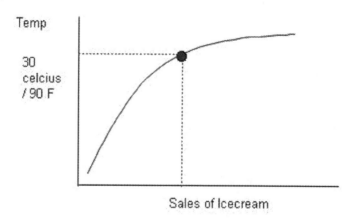

Figure 3-1 Temperature and Icecream

The way this industry utilizes weather was as follows. Basically long term weather forecast was not applied. The sector has a baseline production level, which may be similar to that of the previous year. Upon this, the sector updates weather information. Different from apparel industry, icecream sector is affected by temperature as well as rainfall and humidity. According to industry, they have empirical data on the sales and weather, which they put more confidence than weather bureau's forecast.

Electronics

In this section, an interview was performed at the Hynix semiconductor Inc., one of the major manufacturers of memory chips in the world. As one of the major producer of memory chips, the firm was quite sensitive against fine dust that might affect yield rates of their products. In any semiconductor plants all over the world, similar air quality controlling devices are utilized. Interestingly enough, however, this industry is less affected by weather phenomenon itself. Only eye catching meteorological phenomenon was 'yellow dust' that comes from China each Spring.

Automobile

Automobile, by its nature, is exposed to weather during its usage life.

Automakers, knowing this, try to test heir products. In this research GM Daewoo, a local automaker, was interviewed. Automakers are aware of such meteorological phenomenon as snow, wind, rain, dust, and hot weather in designing and producing cars. In designing cars, depending on final consumer markets where cars would be sold, in tropical areas where weather is all year around summer, heater is not provided. In cold regions, air conditioner may not be provided. Regions where dust causes problems, terms for after service and warranty differ from the same model cars sold elsewhere.

One of the most commonly used fuel, gasoline, also affect design process, since gasoline reacts to temperature sensitively. Temperature also affects assembly process. Under low and bad weather conditions, rate of 'lemon' products can be increased. Hours of sunray also impacts production, logistics, and sales of cars. Generally automakers prefer long sun hours. Also severe storms or other similar weather conditions not only increase delivery related difficulties and costs, but also may cause corrosion of press made parts in inventory.

Cement

This industry showed that while they are reliant on weather information, this is not the most important factor in deciding production volume. Cement, as a product, is sensitive to humidity. In high seasons when humidity is not high, sales volume of cement is high. During low seasons when humidity is high, the number of rainy days is a critical variable. Even though rainy days are limits, the industry believes in yearly average of rainy days in planning production volume.

After humidity, temperature is the second important factor, since construction firms tend to delay jobs when it is too hot or too cold. The industry's top most critical factor is economic cycles. In other words, whether economic cycle is up or down would be the key factor for sales and production, and not weather.

Construction

Construction sector is heavily affected by rain and strong wind, During heavy rainy season, construction works are delayed sometimes by 2 to 3 months. Strong winds like typoons, tornados would have devastating effects on construction sites. Delays mean increased costs, which is critical to construction firms.

Insurance industry

Insurance firms have deep interests in weather information. In many of insurance contracts, such as wind, hail, auto insurance cases, weather is an important factor. Agriculture insurance is heavily affected by meteorological factors. Insurance sector already has statistical database in which the industry has accumulated when and what types of incidents occur most.

Recreational Services

Winter and summer resorts heavily depend on weather. Winter resorts, for example, delay their season opening when average temperature is high. For the industry, temperature and humidity are the two main variables. There are certain points that can be against what ordinary people conceive of the resort industry. Some of the many misunderstood contents are as follows. First, people believe that the more snow the winter resorts have, the better the resorts would be. According to industry, when there is too much snow, the industry is concerned that too much snow may deter people from accessing the ski resorts for their chores to dirve upto the resort area. People may choose to go to resorts where snow accumulation is moderate so that they can drive easily.

Wind also affects the operation of ski resorts in the sense that strong wind may stop the operation of lifts. In the summer, there is no definite relationship between high temperature and the number of customers in resorts, due to other variables. In the summer, among meteorology variables, rain fall affects the customer decisions most, according to industry interviews.

Knowing all these complexities, resorts, by themselves, tend to have their own weather data, while they still want government meteorology bureau to develop more specific weather information processing. Otherwise they may have to turn to private sources.

1)   Climate adaptation cases from Korea industries

Transportation service: Hybrid Bus operation

A local busliner company, named Gimpo Transportation, operates 10 out of its 20 fleet of buses with CNG buses built by the Hyundai Motors Inc. These buses are 30-50% more expensive than conventional diesel powered ones, yet the firm was eyeing on the CNG buses' fuel economy, which would bring benefits over the the time period the buses would be operated.

Separate from the Gimpo Busliner case, the city of Seoul government also launched its effort to introduce electric powered buses in Seoul with Hyundai motors and Daewoo bus Inc. Introducing full electric power only buses will take some time, which has pushed the city to introduce CNG electric buses first. CNG buses are being operated by CNG and battery powered electricity.

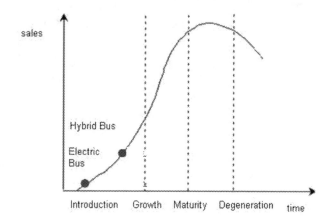

Figure 3-2 Technology Life Cycle: *Transportation service*

In terms of technology life cycle, hybrid bus system is in a relatively later part of the introduction stage, while electric bus is its incipient stage in introduction phase (Anderson, C. et. al. 2004; Boschert 2006).

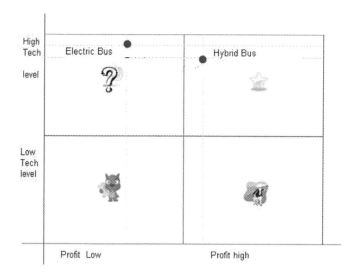

Figure 3-3 BCG matrix: *Transportation service*

From a point of view from manufacturers, hybrid bus can be located in the realm for star business portfolio, since market has acknowledged its entity and profits are expected (Boschert 2006). In comparison, full electric bus is still in development stage(Kirsch 2000; Larminie 2003), and profit level is not able to be evaluated at the moment, which locates itself in the 'question mark' part in BCG matrix.

Transportation service: Korean airlines

Like other major airlines, Korean air is making its efforts to be economicallyCompetitive as well as environmentally friendly. The company has started its efforts since 2003, and eco friendly disposal has been systemized. Fuel saving efforts have been implemented throughout its service areas like other competing ones. For examples, taking economical routes as a starting point, the airline has tried to reduce weights in its airplanes.

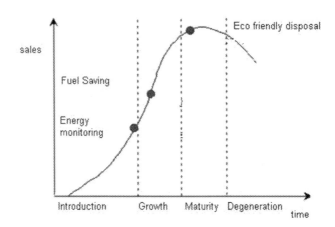

Figure 3-4 Technology Life Cycle: Korean air case as a typical airliner

In disposal side, service items inside aircraft were disposed in eco friendly ways, and all left over food was sent to incinerators. Oil related disposable items were all contracted out to specialized firms to be environmentally safer.

In technology life cycle diagram, eco friendly disposal has reached maturity stage, while fuel saving is in growth stage, followed by energy monitoring in introduction side.

Junmo Kim

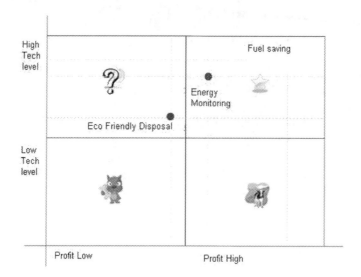

Figure 3-5 BCG Matrix: Korean air case as a typical airliner

In BCG matrix, energy monitoring and fuel saving were located as stars, while disposal part was in 'question mark' area.

Detergent & washers

A local firm, gyungwon enterprise achieved an International standard for no-detergent washer, which was commericialized by Daewoo electrics Inc. This washer uses conventional running water into highly soluable water so that the washing machine does not need degtergent in washing clothes.

In terms of technology life cycle, this machine is in growth stage, while in BCG matrix this machine is located in the area of 'star' which has commercially successful and technological high profiles.

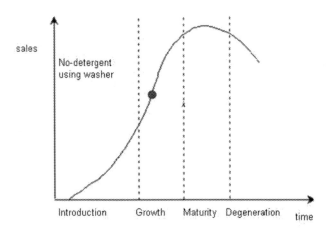

Figure 3-6 Technology Life Cycle: *Detergent & washers*

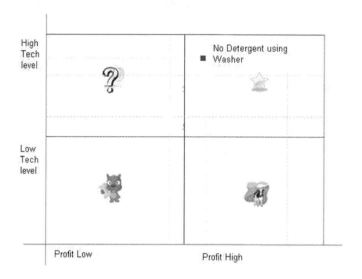

Figure 3-7 BCG matrix: *Detergent & washers*

Utilizing disposable material as fuel

Disposable material can be used as fuel in industries. One example comes from the manufacturing process for cement. The industry is quite well known for burning a lot of fuel in manufacturing process. Ssangyong cement in Korea has been utilizing tires, recyclable plastic, and recycled oil as fuel to produce its cement since 1990s. As of mid 2000s, the firm could reduce about 15% of its coal imports by using the recycled items, which can be translated into 20 million U.S. dollars.

One additional benefit from using recyclable items as fuel is that in this process dioxin is mantled since the temperature inside furnace exceeds 2,000 celcius. Another cement manufacturing firm, Dongyang cement, installed a tele-monitoring system to optimize the reduction of environmentally harzadous material.

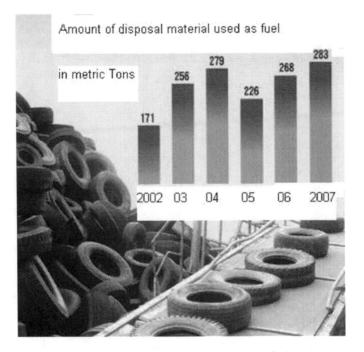

Figure 3-8 Disposal material used as fuel

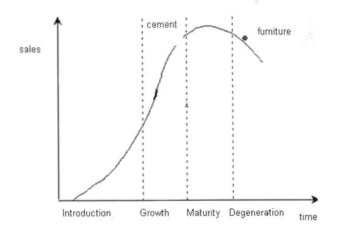

Figure 3-9 Technology Life Cycle: Disposal material used as fuel

Furniture industry has long been practicing its recycling efforts. Livart Inc, has been sending left over wood material of 250 tons per month to a local firm where it was manufactured as recycled ply wood boards.

Junmo Kim

Using recyclable fuel for cement can be expressed as in the growth stage, while for furniture usage, it is in degeneration stage. In BCG matrix, cement case can be classified as a 'star', while furniture case can be understood as a 'dog' portfolio.

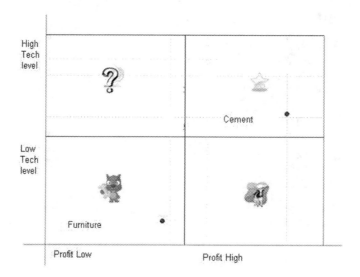

Figure 3-10 BCG matrix: Disposal material used as fuel

Construction: Housing

The epoch when energy saving housing came into a being was the early 1990s when a concept of 'passive housing' was introduced in Germany by a German named Faust.

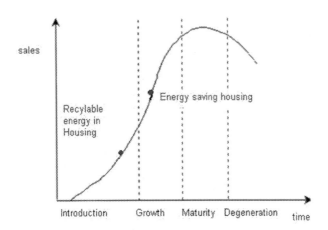

Figure 3-11 Technology Life Cycle: *Construction: Housing*

The core of his idea was to use energy efficient design and material from the design stage and minimize energy costs during use life of a house (Anderson, W. 2006). To accomplish this aim, the design was supposed to use solar energy as a substitute to conventional energy source (Chiras 2006), and suggested that outside wall be 30 centi meter thick so that energy can be preserved and overheating during hot season can be prevented.

Following the trend, Daelim construction co. Ltd. has been working on developing 'eco- 3 liter house' in which heating and cooling costs can be minimized (Salomon 2003; Scheer 2006). An ambitious element of the project is to pursue a zero energy consumption at household as well as to produce electric energy at home and sell it to a power generating company.

Samsung construction company is working on using geo thermal energy in apartment housing units. An example was found in an apartment housing in Daegu area called 'Raemian', in which geo thermal energy saved 17 tons of CO_2 emission annually by using the geo thermal energy source for operating health facilities in the complex. The same company introduced snow melting system for road using geo thermal energy in the same apartment facility. Furthermore, the same company introduced solar power generating system in Yong-in area, producing 76 MWH annually.

In terms of technology life cycle, energy saving housing is in growth stage, while recyclable energy in housing is in its introduction stage (Elliot 2003). In BCG matrix, energy saving housing is located in 'question mark area, while recyclable energy in housing is being located as a cash cow. This can be interpreted as follows. The energy saving housing has been in development for a long time, which means technologically it is established. The only uncertainty with this would be its profitability or market potential. This has located this technology in th question mark area. In comparison, recyclable energy in housing showed its strong gains with examples, which could lead new consumers. This can be translated into being a cash cow.

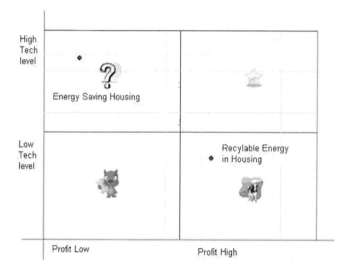

Figure 3-12 BCG matrix: *Construction: Housing*

Retail package

So called the green packaging movement has been quite active in Korea like other economies. Basically all the big retail chains have been participating in this.

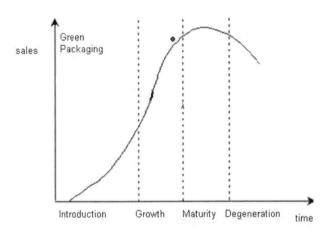

Figure 3-13 Technology Life Cycle: *Retail package*

Among them, the Lotte department Inc. has been practicing with its all 24 stores to reduce packaging in conjunction with its 1,000 apparel suppliers. Hyundai department stores Inc., with its 11 department stores, introduced a 'corn based tray system' to its food sections as a way to cope with environment friendly packaging. There are also other 'smart' cases. Samsung-Tesco introduced a reward system in which if a customer buys 'green milaege' products, the person would get 2% of his or her purchase amount as 'green milege.

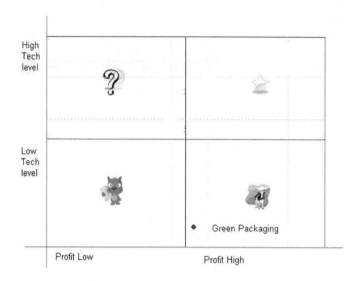

Figure 3-14 BCG matrix: *Retail package*

Other manufacturers of everyday household products including cosmetics, food and sanitary products, are active in reducing traditional packaging and encouraging green packaging. Just listing a few of them would include Yuhan-Kimberley, P&G Korea, Orion, Dongwon F&B, CJ Food, and Daesang.

2. Climate adaptation & Weather utilization cases from the U.S., and European industries

Transportation

Virgin Atlantic airlines together with Boeing and General electric have experimented the use of bio fuel made of coconut and palm oil. In this experiment, a 747 commercial liner was filled with one of its 4 fuel tanks filled with 80% jet fuel with 20% bio fuel of palm and coconut oil. The-experiment was successful, and the firms plan to increase the mix ratio up to 50% in ten years. An implication from this experiment is that bio based fuel can reduce the generation of carbon gas in the sky where planes are flying (Sperling and James 2006, 2004), since at that altitude one pound of fuel burnt can cause three times of cabon gas is emitted to air.

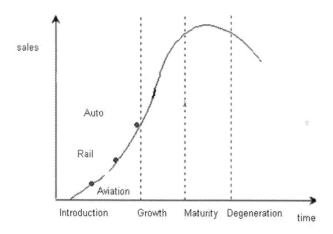

Figure 3-15 Technology Life cycle: *Transportation*

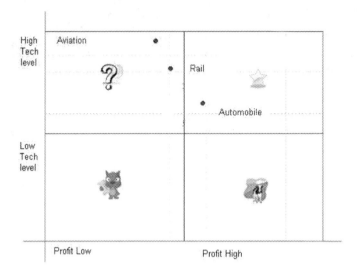

Figure 3-16 BCG matrix: *Transportation*

While using bio fuel is an innovative one, still moderate ways of adapting to climate change are found as well. Air France, like other airlines, by replacing old airplanes with new ones, could reduce CO_2 emission. Furthermore, by using hub airports and by reducing weights of equipment used in cabin, the firm could save money as well as reduce the emission.

As for train, Japan's JR train company introduced a diesel and motor hybrid train, through which energy can be recycled when trains are reducing speed by letting the motor to generate power. In automobile areas, bio diesel has been in the market for some time. EU has a plan to run 10% of vehicles with bio diesel fuel by 2020.

In terms of technology life cycle, transportation field is in introduction stage, while there is a degree of sequence among automobile, air transportation and train in introduction stage. In BCG matrix, aviation and rail are located in 'question mark area, while automobile is located as a cash cow.

Wind Farm

According to Earth Policy Institute, in 2007, total capacity of wind farm in the world was about 59,100 Mega watts, in which Bestas of Denamk took 27.9%, followed by GE wind of 17.7% and Enercon of Germany with 13.2%. Especially Denmark takes 23% of total electricity supply from 5,200 turbines with 3,136 mega watts capacity in 2007 with eventual goal of 0% fossil fuel in 2050 (Mathew 2006).

While EU is planning to implement the 20-20-20% policy, in which EU reduces greenhouse gas by 20% based on 1990 emission level, actualize the ratio of reusable energy by 20%, and increases energy efficiency by 20%, average costs for wind farm was known as about 8 cents per kilo watts, which is still more expensive than coal burning power generation with 5 cents per Kilo watts (EU 2004).

Wind power generation in other industries

A textile firm, Interface, utilizes 16% of its energy demand from reusable energy, and its 7 major plants use electricity from wind, solar and bio energy sources. Starbucks purchased 'wind energy certificate' to provide 20% of electricity needed for its stores in the U.S. and Canada.

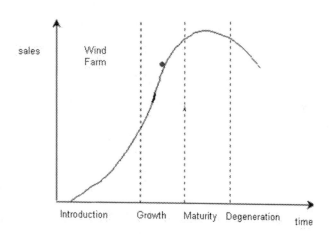

Figure 3-17 Technology Life Cycle: *wind farm*

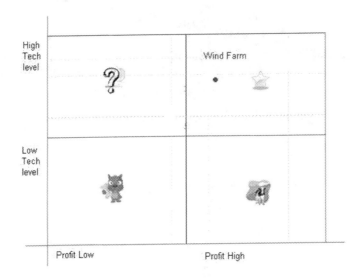

Figure 3-18 BCG matrix: *wind farm*

In terms of technology life cycle, wind farm and power generation is in growth stage, while in BCG matrix, it is located in 'star' area(Mathew 2006; Gipe 1999). As seen from different cases, although average costs are still high, power generation with wind has taken off to be a 'star' in the matrix.

Conventioanl Power generation

Duke Energy is one of the largest power generation company in the U.S.. The firm also supplies natural gas to Ohio and Kentucky. In 2007-2008, approximately, emission of CO_2level recorded was 180 million tons annually. The firm had a goal to reduce the level by 10 million tons by 2015. In addition, the firm's 'Save-a-Watt' program is geared toward increasing efficiency, which will lead to reducing the number of new power plants to be built. In renewable energy area, the firm purchased 100 million watts of energy from wind farms in the state of Indiana.

Electronics

EASTMAN KODAK had emission of CO_2equivalent of 2.86 million tons in 2005. The firm already had reduced its emission of CO_2equivalent by 17% betweem 1997 and 2005, while reducing its energy consumption by 18%. Especially energy consumption was reduced by 12% during 2002- 2005 year time frame. The firm had a goal to reduce its CO_2equivalent emission level by 10% compared to its 2002 emission level.

HP has participated in climate adaptation by reducing its emission of PFCs by 26% in 2003, compared to year 2002. Intel also has shown a progress by reducing energy consumption by 4% each year from 2002 to 2010. From 2001 to 2003, it reduced PFCs emission by 35%.

Chemical

Pharmatheutical company, Bayer, has been working on to reduce its emission level to 53% of 1990 by 2010. Already the firm has reduced its emission level to 65% of 1990 level by 2004. In order to reduce CO_2level, the company replaced coal by gas, which resulted in about 600,000 tons of CO_2emission reduction.

Dupont has been working on to reduce its emission level to 65% of 1990 level by 2010. The firm has increased its production by 35% since 1990, yet its use of energy has increased only by 9% and CO_2emission was achieved by 60%. Similarly, Johnson & Johnson has been working on its emission reduction plan to reduce the emission by 7% from its 1996 level by 2010. The company claims that even though its sales has increased by 372% from 1990 to 2006, its CO_2emission was reduced by 16.8% and at the same time the firm acquired 34,500 tons of emission rights.

Retail

Walmart has been practicing its goal to reduce its emission by 20% and energy consumption by 30% through its 7,000 stores world wide by 2013. It also plans to reduce packaging costs by 5% by 2013, and volume of garbage by 25% (Seth et.al. 2005). It also replaced conventional illumination by LED(Light Emitting Diode) in refrigerators, and adopted solar power generation for shops in Hawaii.

Chapter 4
Policy Choices
& Directions

1.   SWOT analysis of Climate Industry in Korea

In discussing policy choices and directions in this chapter, it is reasonable to start with SWOT analysis to see where current status of climate industry is located in Korea. As one sees in the figure below, Korea is utilizing 43% of its energy source from petroleum, followed by coal of 25% and nuclear power for 15%. By sector, the energy use can be break down into industry usage of 57%, transportation for 21%, and commerce and household use of 20%. In a rough estimation, one can still guess that more than 70% of the energy use is used for economic activities (Elsayed et. al. 2003; Drapcho 2008)

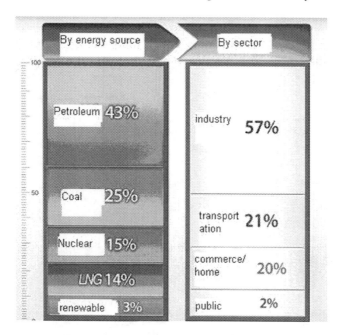

Figure 4-1 Energy composition by sector

Petroleum Usage

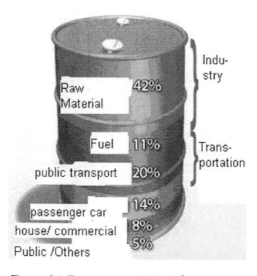

Figure 4-2 Energy consumption by purpose

When petroleum usage is broken down, 42% was used as raw material in production activities and 11% as directly related to transportation in production activities. 20% was used in public transportation, while passenger cars took about 14%. Household and commercial marked 8% of the total petro usage, followed by 5% in public usages and others.

(Strength)	(Weakness)
Abundant infra structure huge growth potential for market Gov't's willingness to support	Highest increase rate of green house gas Manufacutring Orietned economy structure Cumbersome regulation
(Opportunity)	(Threat)
Advent of new business items Financial derivatives Increased awareness of risk management	Management risks increased Increased production costs Becoming a trade issue

Table 4-1 SWOT analysis for climate industry Strength

On strength side, Korean economy has abundant infra structure from which climate industry can grow, since climate industry can grow from the industrial basis for conventional industrial sectors.

Huge growth potential can be evidenced by economic indicators. In nominal GDP, Korean economy was seated on the 13th, while trade statistics shows that Koera has the 11th largest trade volume in the world. This shows that the economy has necessary conditions to have seizable size of climate industry.

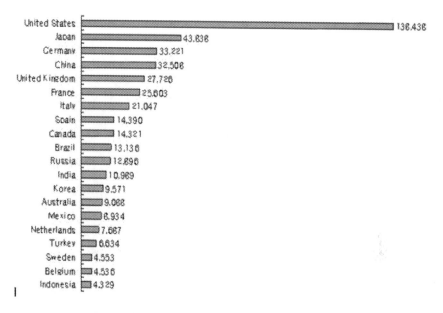

Figure 4-3 Nominal GDP of major nations 2007

Source: IMF statistics

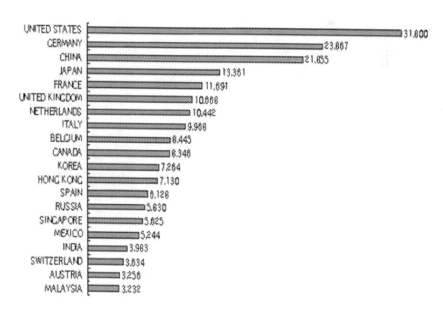

Figure 4-4 Trade volume of major economies, 2007

Source: IMF, Korea asso. of Trade

Weakness

In terms of annual average increase rate of Green house gas by nations for the period between 1990-2001, Korean economy showed the highest increase rate. This is clearly weakness as well as a necessary condition to have strong climate change industry.

When comparing total emission amount, Korea marked the 9th among nations. In 1990, Korea's emission rank was 12th with 263.223 Gg CO2e, which has changed into 9th rank with 508,252 Gg CO2e in 2001. Among Annex I countries, Korea's rank was 10th, in which the U.S.A. was at the 1st rank in terms of emission. This fact that Korea's absolute amount of green house gas emission is high and at the same time has the highest growth rate would clearly sigal that costs to deal with the green house issue will be hefty for firms (Labatt 2007; Lecocq 2005).

Korea's relatively high dependence on manufacturing also becomes a weakness factor when discussing weakness in SWOT perspective. As one sees in figure , proportion of manufacturing sector in GDP has been maintained between 24 to 27% through the years after year 2000. These are clearly low numbers, compared to the country's historical data. Yet, it is still high in a comparative sense, which positions the country in weak position in terms of climate change adaptation (Burroughs 2001).

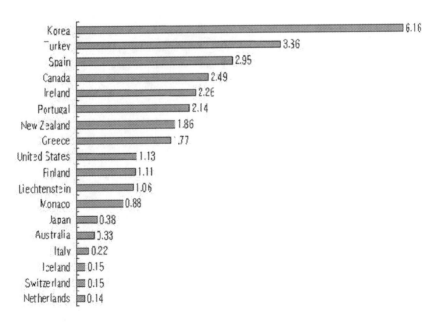

Figure 4-5 Annual average increase rate of Green house gas by nations, 1990-2001

Source: UNFCCC.

Country	1990	2001	Country	1990	2001
USA	5,529,241 (1st rank)	6,259,501 (1st rank)	IRELAND	55,495 (28th rank)	70,942 (22nd rank)
RUSSIA	3,166,421 (2nd rank)	2,282,531 (2nd rank)	DENMARK	70,993 (23rd rank)	70,468 (23rd rank)
JAPAN	1,179,935 (4th rank 2)	1,229,731 (3rd rank)	AUSTRIA	67,151 (26th rank)	65,947 (24th rank)
GERMANY	1,199,619 (3rd rank 3)	1,002,029 (4th rank)	BULGARIA	110,692 (19th rank)	58,276 (25th rank)
UK	774,310 (6thrank 6)	676,445 (5th rank)	FINLAND	49,610 (30th rank)	56,018 (26th rank)
CANADA	473,310 (9th rank)	620,582 (6th rank)	NEW ZEALAND	42,920 (31st rank)	52,541 (27th rank)
FRANCE	533,314 (7th rank)	521,477 (7th rank)	SWISS	51,045 (29th rank)	51,893 (28th rank)
AUSTRAILIA	499,903 (8th rank)	518,287 (8th rank)	SLOVAKIA	69,662 (24th rank)	45,438 (29th rank)
KOREA	263,223 (12th rank)	508,252 (9th rank)	SWEDEN	68,652 (25th rank)	42,231 (30th rank)
ITALY	437,033 (11th rank)	447,792 (10th rank)	BELARUS	105,333 (21st rank)	40,985 (31st rank)
POLAND	452,685 (10th rank)	371,041 (11th rank)	NORWAY	35,032 (33rd rank)	27,443 (32nd rank)
SPAIN	244,603 (13th rank)	336,680 (12th rank)	CROTIA	25,271 (35th rank)	18,094 (33rd rank)
UKRAINE	872,377 (5th rank)	329,323 (13th rank)	SLOVANIA	15,351 (36th rank)	14,443 (34th rank)
NETHERLAND	215,355 (14th rank)	218,594 (14th rank)	LITHUANIA	38,631 (32nd rank)	11,875 (35th rank)
TURKEY	126,527 (18th rank)	191,922 (15th rank)	ESTONIA	33,262 (34th rank)	11,166 (36th rank)
BELGIUM	144,335 (17th rank)	144,139 (16th rank)	LUXEMBURG	12,413 (37th rank)	9,557 (37th rank)
CZECH REPUBLIC	194,493 (16th rank)	142,380 (17th rank)	ICELAND	5,442 (39th rank)	5,534 (38th rank)
GREECE	105,549 (20th rank)	127,990 (18th rank)	LIECHTENSTEIN	223 (40th rank)	250 (39th rank)
ROMANIA	212,887 (15th rank)	103,693 (19th rank)	MONACO	107 (41st rank)	118 (40th rank)
PORTUGAL	63,749 (27th rank)	80,449 (20rank)	LATVIA	5,772 (38th rank)	-3,445 (41rank)
HUNGRAY	94,230 (22nd rank)	75,599 (21st rank)			

Table 4-2 Green house gas emission by country (CO_2 equivalent)

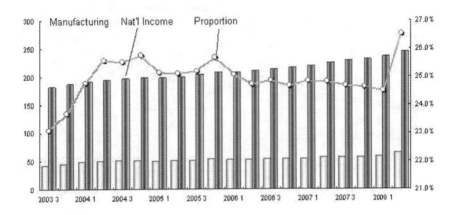

Figure 4-6 Proportion of manufacturing in GDP

Source: Bank of Korea

As another weakness, it is possible to list degrees of government regulation. In the table below, Korea marked 55th in 2005, and 36th in 2007 in terms of firm competitiveness on government regulation policy, on which bigger scores mean more competitiveness. The reason that this is regarded as weakness comes from a point that climate change adaptation would increase firms' costs in addition to their existing burden from government regulations. If firms with relatively high burden of regulation attempts to launch climate change adaptation, it would definitely add up to a high level (McKibben 2007).

	Korea	Singapore	Japan	Taiwan	USA	Germany	China
2005	4.22(55)	6.86(5)	5.68(21)	5.78(18)	5.64(23)	4.29(43)	5.56(26)
2007	4.53(36)	8.00(1)	5.28(26)	5.39(24)	5.66(21)	5.11(29)	6.38(11)

**Table 4-3 Relative comparison of strength of Business
competitiveness vis-à-vis government regulation**

	Korea	Singapore	Japan	Taiwan	USA	Germany	China
2005	4.57(18)	8.62(2)	5.65(30)	5.86(13)	8.22(4)	4.64(43)	5.49(34)
2007	4.22(36)	8.85(1)	5.64(22)	7.56(8)	5.33(24)	4.37(34)	6.62(7)

Table 4-4 Easiness of business operation

A similar measure is 'easiness of firms' activities shown in table .This will act as a barrier for firms in climate change adaptation.

Opportunity

The advent of new emerging industrial and service sector industries clearly work as opportunity for the growth of climate change industries in Korea. While lists and examples of those industries were exemplified in previous sections, some of the details for boundaries of the new sectors can be highlighted. These new sectors can be a highly sophisticated manufacturing sectors that can contribute to the development of climate industries (Nordhaus 2003, 1998). Second, the new sectors may include advanced service sector firms like financial institutions. Third, a mix of service and manufacturing style firms can also be included (McKibben 2007).

Awareness of firms and individuals on the necessity to manage and address risk management can work as opportunities for climate change adaptation and its related industries (Esty and Winston 2006; Botterill & Wilhite 2006; Daniels et. al. 2006).

Development of financial institutions and its working can expand the realm of climate industries, which have been witnessed by the advent of future markets for commodity and agricultural products (Ruth 2006; Guha 2006).

Threats

Growth of climate industry also has its threats. Above all, most CEOs regard climate change as an increase of management costs, which they feel as burdensome (Hart 2007).

Climate change adaptation very naturally increases production costs involved, which industries would not like (Haurie and Viguier 2006).

Increased awareness on environment may also increase potential sources for trade conflicts among countries and regions.

2. Policy Alternatives for climate change adaptation

1) Micro level policy

Micro level policy denotes an array of policy measures that can be designed and used to promote climate change adaptation by promoting industries or technologies at firm level, which is sharply contrasted with macro policy to be continued in the following section through which a nation level or international regime level policy and its supporting policy mechanism. Micro level policy is typically a promotional policy for technology development, which is not foreign to policy community in industrial and technology policy development (Kim, J. 2002, 2005a).

'Star' business support policy

This policy idea suggests supporting either manufacturing or service sector firms that have recent 3 year average profit ratio of 5% or higher and maintain technology gap between them and advanced peer firms in other countries less than 4 years.

The main purpose of this project or policy would be to reduce technology gap between them and the advanced peers from 48 months to 18 months after the promotion.

The policy idea suggests that it supports selected manufacturing and service firms for 3 years, while the amount shall be left for a committee to be formed. Also there should be some consideration to weigh either of manufacturing or service sector over the other in deciding the amounts.

For the firms that received support, technology level and economics of technology review should be followed periodically.

Examples of star business items

Based on the CEO survey in this research, the following items were suggested as examples for the 'star' business items.

Fuel cell powered vehicle / Transmission for fuel cell vehicles / Electric Scooter/ Battery powered vehicles, fuel cell automotives (Boschert 2006)

In this segment of market, the key issue would be to increase efficiency in battery so that economies of scale can be achieved for industries and individuals.

Vehicle emission reduction technology

This type of technology and products are hot in demand without doubt. Key to this item would be to improve over the existing models and technology used in the market.

High efficiency heat pipe / Heat recycling system / Energy saving heater /

These items are all related to heating and airconditioning in buildings, which have huge market potentials.

Improving solar collector

While solar panels have been around for several decades, it has always been challenged by low benefit to costs ratio and low efficiency, which means improving the problems are one of the imminent problems to improve solar based power generation.

Middle & Long term promotion items: 'Cow ' business support policy

This policy aims at supporting industries and technology items with recent 3 year's profit rate of 3% or higher and technology gap vis-à-vis advanced firms in other countries greater than 4 years.

Therefore, policy objective would be to reduce the technology gap from more than 4 years to within 36 months after promotional period.

Business and technology items under this category would be understood as those sectors that can maintain their visibility, yet, with their technology gap against advanced firms in the sectors, have limited chances of long term competitiveness.

Examples of 'cow' business items

Financial derivatives for agricultural products / Financial derivatives for raw material

Finance sector can be benefited from climate and weather. One critical hurdle to overcome would be how to persuade people to subscribe to their products. This would be a long term project that should go with the growth of economies in a complex way. Another caveat in this would be over confidence in derivatives.

Hybrid vehicle/ fuel cell engines & parts

Hybrid cars and their related technologies are objects of long term updating and development. Some of the technologies can fall into 'star' business category, while the others can be classified as 'cow' segement.

Waste treatment business

This segement would also be a promising element, yet requirements from society tend to grow as time passes by, and this would pose a continuing challenge to the industrial sector (Imhoff 2005).

Body temperature control apparel / Ultra light high performance textile / 100% recyclable paper based clothes

High performance clothing and apparels would be one way to adapt to climate change. The new changes make apparel and textile sectors high tech sectors.

Meteorology Equipment/ weather consulting

This segment clearly shows a trait of a oligopoly due to limited market size and quality required. As a long term goal, countries may feel necessary to develop their own equipment. Weather consulting, same as climate based financial products, needs time to persuade people to let them actually purchase consulting products. This makes the sector a candidate for a long term promotion.

Next Generation Business support policy 1: 'Question mark' items

In this category, policy criteria can be summarized as follows.

This policy aims at supporting industries and technology items with recent 3 year's profit rate of 5% or lower and technology gap vis-à-vis advanced firms in other countries less than 4 years.

The meaning to have this segement is that there are some cluster of industries whose technology gap vis-à-vis the top notch firms are not that big, yet their profit performance has not been good enough to make catch-up efforts.

2) Macro level policy

As mentioned in the above, macro level policy should be synchronized with movements in international arena by its nature. Any national level efforts in macro level plain field can not override basic policy directions set up by international regime like the Kyoto Protocal and others.

Carbon Market Policy

One of the typical policy alternative for a national government in macro policy arena is the carbon market policy (Trabalka 2005; Tang 2007, 2005; Swingland 2003; Yamin 2003). Like any other policy cases, if and only if carbon market can be generated fully without government's activities, it would be more than desirable (Bayon et. Al. 2007; Bond et. Al. 2009). This may not work like in many other policy cases, which brings some necessity for governments to enter into the area (Tietenberg 2006; Snow 2007;Soon 2001).

World Carbon Market Size		2006		2007	
2006 / 2007 COMPARISON		Trade Volume in hundred million tons /	Trade Amount in hundred million US dollars	Trade volume	Trade amount
Emission Rights (AAU)	EU-ETS	11	244.4	20.6	501
	OTHERS	0.3	2.6	0.5	3.0
	SUB TOTAL	11.3	247	21.1	504
Project Market	CDM	5.6	62.5	7.9	128.8
	JI	0.2	1.4	0.4	5.0
	OTHERS	0.3	1.5	0.4	2.7
	SUB TOTAL	6.1	65.4	8.7	136.5
	TOTAL	17.4	312.4	29.8	640.5

* WB, State and Trend of the Carbon Market 2008

Figure 4-7 World Carbon Market size

As shown in the figure, world carbon market is already growing fast, which calls for both private and public attention (Lecocq 2005; Lynas 2007; Griffiths 2005; Kojima 1998).

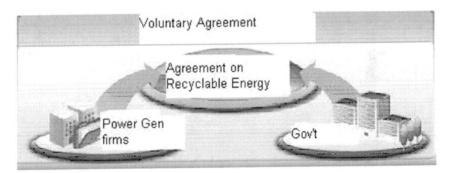

Figure 4-8 Voluntary Agreement

Green Energy Policy

As part of macro policy, green energy policy can be sketched out. Under this grand scheme, a set of areas can be covered (Wigley 2005; Walz 2009; World Bank 2008). In ecology, policy should address to prevent the extinction of species. One of the ways would be to secure DNA of the endangered species(Brower 2000; Cipiti 2007; Firor 2002; Grabher 1994, 2001; Hawken 2007). As for the water problem, which is a common issue all over the world, policy should focus on providing a master plan for water supply and demand (Langley and Curtis 2004).

From air and forest to agriculture and public health, macro level national policy should address the issue by defining the problems, setting agendas, and prepare a basic policy in harmony with other related areas in line with climate change adaptation (Maracchi et. Al. 2005; Mannion 2006; Maddison 2001; Moser 2007).

Sector	Impact	Policy
Ecology	- Extinction of species - New specifies may move from other regions	- Prevent the extinction / secure DNA of endangered ones. - Cure ecology to restore by itself
Water resource	- Shortage of drinking water - Shortage of argricultural, industrial use water	- develop water resources - master plans in line with climate change
Air	- Frequent Ozone warning - Yellow sand wind & Increase of desert area	- Ozone warning system - Support for 'Clean car' production
Forest	- Forest fire and other disasters increased - Increase of diseases	- Danger detection system - Securing equipment
Agri-Culture	Damage from disease, insects, weather factors and climate	- Genetic engineering for weather resistant seeds - Disaster reduction system
Cattles	- Climate change and adaptation - Spread of new diseases	- Improving with genetic engineering - Animal disease control system
Public Health	- Hot weather and its impact - Epidemic disease prevention	- Hot weather warning system - Disease control system
Energy	- Depletion of fossil fuel	- New energy sources
Industry	- Changing environments and adaptation to them	- Changing industrial landscape

Source: Ministry of Environmental affairs National policy for climate change preparation, 2007.

Table 4-5 Summary of Policy in Korea

Roadmaps

Based on empirical surveys introduced in this book and implications from research can be summarized in a series of policy road map type presentation for a few selected sectors or technology groups.

Automobiles / related machinery

As this book has reviewed and analyzed, automobile industry has been already going through a transition into hybrid and fuel cell regime, which will be accelerated in times to come. Not only for policy making purposes, but also for private sectors' point of view, a possible road map per se would be to insure a whole range of hybrid, battery and fuel cell type energy efficient and climate change adapting technology and products are lined up in time horizon (Trabalka 2005; Ruth 2006Toman 2001;).

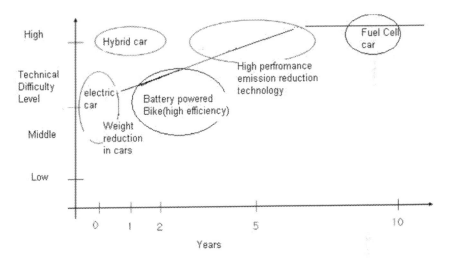

Figure 4-9 Road Map: *Automobiles / related machinery*

Services

Service sector industries range from a simple and low end ones to high capacity and yielding consulting services. In this research, service firms included major airlines, which are shown in the roadmap. Regarding climate change, high flyer services very naturally include financial services and consulting, while weather insurance is regarded as technologically demanding than higher level consulting operations (Benson 2004; Bassam 1998; Asafu-Adjaye 2005). If a consulting industry is to support climate change adaptation, this road map can be an example where the industry should be headed on.

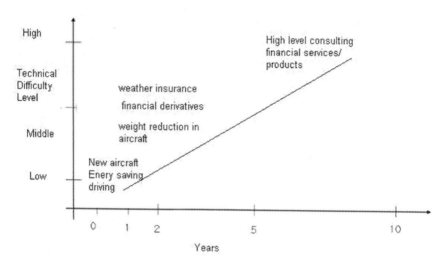

Figure 4-10 Road Map: *Services*

Energy / Construction

Energy sector is one of the sectors where the range of technology level is so diverse. One of the simpliest way to adapt to climate change by the sector would be to work on recycling energy, followed by increasing efficiency in recycling equipment and heat pipes (Scheer 2004; Smith 2004, 2001; Strong 1993).

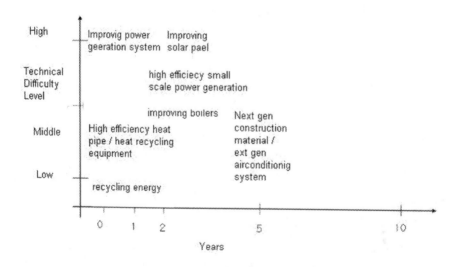

Figure 4-11 Road Map: *Energy / Construction*

Eventually, energy sector's contribution will come from improving power generation system including solar panels (Bradford 2006) and small scale efficient power generation. Material side like construction material also takes an important pillar (Yannas 2005; Strong 1993; Hyde 2007; Boyle 2004) .

IT

IT industry has a characteristic that can assist other sectors. IT field can help energy areas with fuel cells and battery, while material areas can be assisted like metal masks. Faithful to the basics of the sector, software development can also contribute to climate change adaptation.

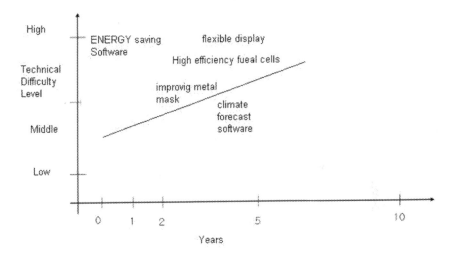

Figure 4-12 Road Map: *IT*

3. Policy directions for promoting weather industry

While included within the purview of climate industry, weather industry is a distinct subset of the climate industry. Because it has its distinctive nature, policy directions for this industrial segment sometimes have different tastes (Flannery 2005; Kurokawa 2007). In this section, a comprehensive SWOT analysis will be presented to address policy directions for weather industry with possible policy directions.

As for general background of the weather industry as a whole, the following SWOT analysis can be presented. As for strength, private weather firms have a long standing relationship with the Korea Meteorology Administration

(KMA) together with high recognition as industrial players in government. Weaknesses would include a scale issue, which is far below economies of scale, and 'barren' market conditions. Furthermore, relationship with government's KMA would be a weakness factor as well. The reason being that if private firms began starting commercial business, KMA's existing favor would be evaporated.

Internal Environments / Outer Environments	Strengths -Long relationship with gov't -High accessibility to weather market -High degrees of understanding on gov't policy	Weaknesses - Difficulty of initial market -Small scale -Relationship with gov't threantened
Opportunities -Potentials for growth -Gov't interests for promotion -Low entry barriers	SO Strategy	WO Strategy
Threats -Difficulty of paid services	ST Strategy	WT Strategy

Table 4- 6 SWOT from a realistic view point

As regards to opportunities, it is possible to look at huge growth potentials together with government's willingness to promote the industry. Also low entry barrier would be an opportunity factor (Flannery 2006; Johansen 2002, 2006). For most people, it would be hard to understand why government policy may have multiple façade. In this case, if private firms do enter into weather market, KMA will feel uneasy about the firm, since they would become rivals. At the same time, government as a whole as well as the KMA has plans to promote weather industry. How can one qualitatively differ these statements? To this, it is possible to argue that even within the government, there could be different ideas, and diversity of ideas can be even wider when different departments are concerned.

Weather consulting industry

Weather consulting is an area that seems to be easy to start. Indeed, in many countries, private weather consulting firms have been in existence more than two decades. Despite the fact that they are in business, it is clearly another matter whether they are profitable or not. As this book has reviewed in the survey part in chapter 2, one of the critical weaknesses is that market is not sufficiently facilitated (Zillman 1999).

If one investigates into deeper aspects beneath the skin of what we are looking at, the following SWOT analysis can be presented. As for strengths, the firms have know-how on what to do with weather forecast. They are usually composed of expert meteorologists and sometimes added by management specialists who can manage their clients.

While they know what to do, their weakness side is also critical. First of all, Private weather companies have not hit the market. They lacked essential marketing tools to penetrate into the market. But what is more important is that all the weaknesses and problems are all inter-wound that one can not blame the industry itself.

Their second weakness comes from their small scale. They are, in many countries, composed of only small number of people, except for a big firm like the Weather Channel in the U.S.. This small scale goes together with the lack of marketing strategy and power to persuasion against their clients.

Third, they have not overcome, in some countries, the nature of science. Meteorology is science, but doing business does not mean that they maintain the attitude as doing science. Business requires more room in discussion.

Outer Environmnets \ Internal Environments	Strengths -Accumulation of know-how	Weaknesses -Small scale of consulting firms -Lack of marketing capability -Limited consulting technique
Opportunities -Increased awareness of weather marketing - Increased awareness of risk avoidance	SO Strategy	WO Strategy
Threats -Limited trust on consulting firms - Limits from gov't support	ST Strategy	WT Strategy

Table 4-7 SWOT analysis of weather consulting industry

As explained private firms do have limitations, yet they have bright future as well. As seen from the results from the survey in this book, more and more firms are aware of risks associated with weather and climate. These firms want to avoid or reduce the associated risks and costs. They want to find new ways to cope with weather, which were once regarded as unavoidable destiny. Not only producers, but also financial service institutions have opened their eyes to new business opportunities. On the other side, as threats, the consulting industry is exposed to a fundamental issue, reliability. This could be linked to pricing for the service, which in turn, is connected to marketing, overcoming hangovers from science and all other issues.

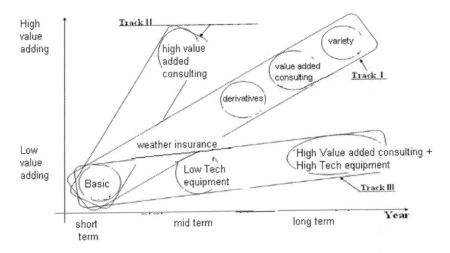

Figure 4-13 Road Map: *Weather consulting*

Then, what could be the possible tracks for weather consulting firms? On this, it is possible to suggest three distinct tracks. The first would be starting with basic weather marketing (Botterill ed. 2006; Cluassen 2001), followed by deepening into weather insurance and financial derivatives in time sequence and eventually leading to value added consulting. Track 2 would be sharing the same starting point as the basic marketing with an end point of high value added consulting. Track 3 is a distinct approach in that it mixes equipment manufacturing with consulting.

Meteorology Equipment Industry

Meteorology needs machines and equipment. As for industry promotion, it is not that easy to have this sector, for several reasons. First, 'economies of scale' for this market is very limited. Of course, similar industrial sector with precision capability can always enter the market, yet this would also incur costs to the incoming firm.

Secondly, for most end users of the equipment of this type, if maintenance is neglected, it would be linked to serious quality problem for the data they produce. Based on industry practice, most government meteorogloy agencies tend to maintain their equipment, while para public agencies, in some cases, tend to neglect maintenance due to budget related issues. At any rate, the end results from para public organizations are usually lowered perception on reliability of the machines they use. This factor would lead to a hurdle in expanding the market.

Third, compared to the case where any type of standardization is established, if there were no equivalent standards, it is quite difficult to expand market due to compatibility problems.

A national champion case of this sector is the Vaisala of Finland, which is not only the champion of the country, but also, due to government support policy, this firm became the world number one supplier of meteorology equipment.

Internal Environments / Outer Environments	Strengths Customization Maintenance capability	Weaknesses - Limited economies of scale - Lack of maintenance - Lack of standardization
Opportunities - Potentials for new markets	SO Strategy	WO Strategy
Threats - Limited demand	ST Strategy	WT Strategy

Table 4- 8 SWOT analysis of meteorology equipment manufacturing industry

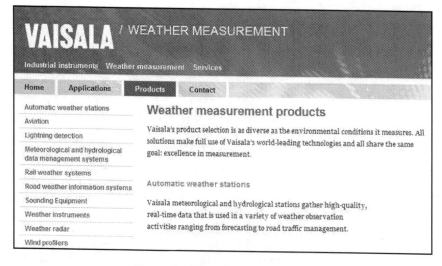

Figure 4-14 Vaisala as an example

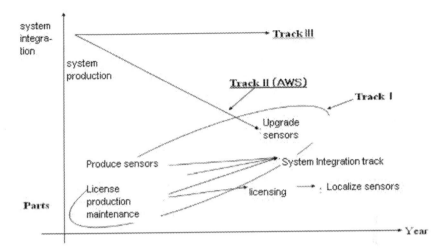

Figure 4-15 Road Map: *meteorology equipment manufacturing industry*

For equipment sector, it is possible to present three tracks for promotion. Track 1 starts with license production and maintenance, and expands into producing sensors and system integration. Track 2 begins with system integration first, followed by parts production. In comparison, track 3 constantly stays with system integration only.

Weather Insurance: Agricultural Insurance

For weather insurance industry, strength factor would be the introduction of agricultural insurance system in Korea, mainly based on private insurance plus some portion for government intervention that has been pre-existed prior to the introduction of private insurance. In contrast, the lack of accumulated know-how of the industry (Daufresne et.al. 2004) would be an important weakness. As for opportunities, volatile weather conditions and growing trend of global warming are 'helpers' of the industry (Doering ed. 2002; Dirnböck et. Al. 2003; Davidson 2005; Cowie 2007; Crick & Sparks 1999; Dauncet 2001), while limited demand and difficulty of forecasting demand for service are so called threats in SWOT analysis.

Internal Environment Outer environments	Strengths - Introduction of - agricultural insurance	Weaknesses - Lack of know-how
Opportunities - Volatile weather - Global warming	SO Strategy	WO Strategy
Threats - Limited demand - Difficulty to forecast demand	ST Strategy	WT Strategy

Table 4- 9 SWOT analysis of weather finance industry

Regarding agricultural insurance, by far, the U.S. case has been the model with its long history and experiences. It would be not wrong to say that Korean introduction of agricultural insurance also has been affected by the U.S. cases, although institutional arrangements are quite different. If private sector portion and subscription rate are to be increased, the two systems may converge in the future.

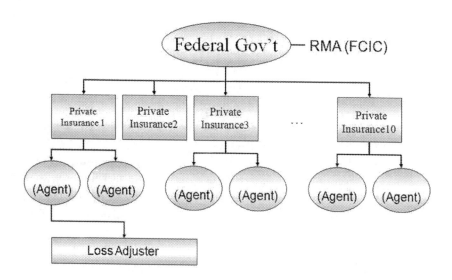

Figure 4-16 U.S. Agriculture Insurance

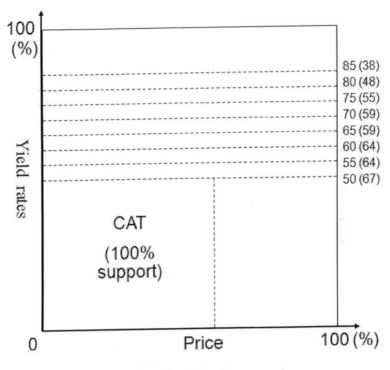

* CAT : full gov't subsidy(free)
* ( ) numbers are % of support

Figure 4-17 U.S. Agriculture Insurance 2

CHAPTER 5
CONCLUSION

Through this book, empirical data based presentation of industries has been discussed. This book intended to show how 'real world' industries perceive and adapt to changing environments of climate change and what would be their next move in terms of technology item based on CEO survey. As shown, this book has been relying on two major surveys to present on commercializing weather and climate change.

As a concluding chapter, this chapter will sum up the discussion, and lay out fundamental issues, and then wrap with a final word.

Adaptation vs. reduction

While implementing survey which was utilized in this book, a common misunderstanding among people could be detected. Among business leaders and even among bureaucrats, it is a well ignored notion to distinguish climate change adaptation and climate change adapting reduction. Of course climate change adaptation can include reduction, yet in this book discussion was focused on a narrow bound of climate change adaptation. In discussing adaptation in the context of climate change, it is possible to think of three sub concepts. The first one is Anticipatory adaptation (proactive adaptation), which takes place before impacts of climate change are observed. The second category is the Autonomous adaptation (spontaneous adaptation), which does not constitute a conscious response to climatic stimuli but is triggered by ecological changes in natural systems and by market or welfare changes in human systems. The third one is 'Planned adaptation', which is the result of a deliberate policy decision, based on an awareness that conditions have changed or are about to change and that action is required to return to, maintain, or achieve a desired state.

<u>Anticipatory adaptation</u> (proactive adaptation)
- Takes place before impacts of climate change are observed

<u>Autonomous adaptation</u> (spontaneous adaptation)
- Do not constitute a conscious response to climatic stimuli but is triggered by ecological changes in natural systems and by market or welfare changes in human systems.

<u>Planned adaptation</u>
- Adaptation that is the result of a deliberate policy decision, based on an awareness that conditions have changed or are about to change and that action is required to return to, maintain, or achieve a desired state.

What this book intended was not to emphasize the planned adaptation. This book originally had an aim to present 'as it is' of the current façade of industries in front of climate change as a phenomenon. Of course, policy directions were suggested in the book, and this clearly implies that 'planned adaptation' part is presented in the book. This part would be one of the two aims of the book. In other words, this book had an equal weight for both describing current façade and prescriptions for the future. In prescription part, this book tried to distill results and implications from the surveys and patterns of technology development.

Then how can one set a relationship between reduction and adaptation? In the figure below, it is possible to show how both concepts are connected. If reduction is geared toward lowering the level of emission, adaptation may be projecting more than the level one can perceive by reduction. Adaptation would mean to change and transform what we have been doing as 'inertia' or to change a paradigm if you would.

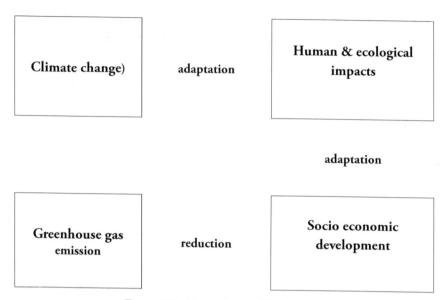

Figure 5-1 Adaptation and Reduction

Costs of Adaptation

Costs occur when adaptation takes place. This is not an exception for climate change adaptation. Line cost1denotes total costs resulting from a status where no global action to cope with climate change is implemented. Cost3 shows costs that occur even in the case where all human efforts are made. In comparison, cost 2 means the sum of cost3, which is inescapable, and the cost of adaptation.

Thus, gross or total benefit of adaptation can be expressed as the difference between cost1 and cost3. In comparison, net benefit of adaptation is the difference between cost1 and cost2. Total cost of climate change after adaptation is expressed as cost2 when cost of adaptation is the difference between cost2 and cost3.

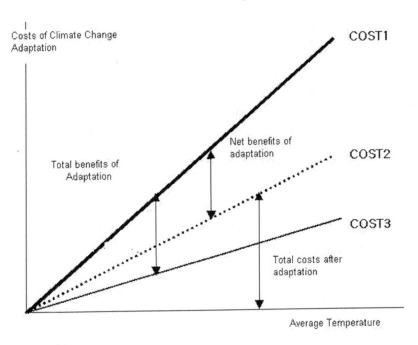

Figure 5-2 Costs of Adaptation

Stern Review: The Economics of Climate Change) p.405

Types of adaptation	Climate status	
	(existing climate; C_0)	
(adaptation to existing climate; A_0)	Adaptation to current climate (C_0, A_0)	Adaptation to current climate at changed climate (C_1, A_0)
(adaptation to altered climate; A_1)	Adaptation to changed climate at current climate (C_0, A_1)	Adaptation to changed climate at changed climate (C_1, A_1)

Table 5-1 Types of Adaptation

Country examples

There are good example countries to introduce, mainly European countries. In this section, only two of them will be presented for the purpose of wraping up the book. In terms of goals, Finland states "To reduce negative impacts of climate change and utilize it as an opportunity", while U.K. policy goal is to grasp major risks and opportunities related to climate change and to set agendas for public private cooperation". On the contents side, Finish policy aims at the following.

- Acquire knowledge on direct and indirect impacts of climate change

- Prioritize sectors with urgent adapatation needs

-To know trends in international communities

In comparison, U.K. policy focuses on the contents below.

- Prepare National level agendas and framework

- Examine existing adaptation activities

- To consider policy designs for the future including expanding incentive schemes

	Finland	U.K.
Goal of Adaptation Policy	- To reduce negative impacts of climate change and utilize it as an opportunity	- To grasp major risks and opportunities related to climate change and to set agendas for public private cooperation
Contents of the Adaptation Policy	- Acquire knowledge on direct and indirect impacts of climate change - Prioritize sectors with urgent adapatation needs - To know trends in international comunities	- Prepare National level agendas and framework - Examine existing adaptation activities - To consider policy designs for the future including expanding incentive schemes
Main Player in Gov't	(Ministry of Agriculture and Forestry)	(Department of Environment, Food, and Rural Affairs)
Sectors under Focus of Policy	Sectoral approach (15 sectors): agriculture, forest, fishery, deer preservation, hunting preservation, industry, water, bio diversity, energy, transportation, land use, buildings, health, tourism, insurance	Sectoral approach agriculture, horticulture/ forest, bio diversity, marine preservation/ fishery, flood control, water resources, energy, architecture, land use, transportation, manufacturing, financial and insurance, retailing, service, leisure / tourism, emergency, health, defense

source: Ministry of Environmental Affairs, National adaptation policy guideline 2007

Table 5-2 Finland and U.K. comparison

Overall, it is fair to argue that in most european countries, policy goals, agendas, contents, and sectors under concern have converged, which shows that there is a growing common ground not only international communities, but also at national level.

Concluding Remarks

Through chapters, this book has discussed how industries perceive climate change from survey results and projected on possible technology fields to be introduced with climate change adaptation backgrounds. As a concluding and closing remarks, it is sensible to start with the 'sociology' of climate change.

Whenever one faces the word climate change, it has different façades (Kushner 2009; Leroux 2005; Pearce 2007). Some comes from science with its pure pursuit, while some science based approach may hear like political rhetoric (Sinnot-Armstrong 2005). Also one finds exaggeration on both sides those worrying climate change and those who would deny climate change. But, as a policy discourse and actual policy agenda, there should be a common ground from which one can honestly start thinking what to do for the future.

That starts from admitting, in some sense, limited scopes from science as well (Carraro 2000; Dessler 2006, 2005). This does not mean to lower the value of scientific research, yet it is also to be alert in utilizing the findings from scientific research. The rational in taking this position is simple. Once we set up and agree on an agenda based on scientific knowledge and finding, its consequences may be irreversible or would cost enormous amounts before we cancel ad go back to where we were.

Global climate change is a huge trend, and it is approaching (Daniels 2006; Dawson 2009; Doering 2002). Yet, limited human capability makes it still difficult to predict the timing of its coming and knowing the just exact time frame to start for preparation. If one can endure more than 100 years, which is longer than average life expectancy, long term would mean for him or her. Yet, human matters do not wait for long term. From this point, the gap of perception, between supporters of science plus policy groups who believe science strongly and people who are realists, starts (Dryzek 2005; drake 2000). Industry people may be the realists per se. For them, it is in some sense meaningless to argue for the upcoming climate change unless they are determined by persuation.

As one famous economist mentioned, "there is no long run". Long run means those who are on earth are dead. This is the subtle point of climate change adaptation. It is not to say that climate change adaptation is not necessary or to reduce the value adaptation related activities. Rather, this perception, by realists, do signal us that there should be clear stages of action plans to follow in the sense that purely science based warnings may reduce our growth potentials if we faithfully follow them. In this sense, the surveys introduced in this book gave us a good example of how business community considers the change and how they may be changed.

Adapting to climate change requires an array of huge changes. Perhaps the most fundamental change might come from the change of our minds. In this pursuit of mind set change and re-orientation, this book believes that there should be a fine balance between science based arguments and the realists' view in approaching the phenomenon. With this book, I hope a common 'playground' from which a more realistic dialogue between the two sides can be prepared.

Appendix

Effects of climate change in Europe

Climate change in 21st century, with global temperature rise of 2,9°C:

- Retreating glaciers, longer growing seasons, shift of species ranges and health effects due to heat waves
- Magnification of regional differences in resources and assets, such as coastal flooding and erosion
- Extensive species losses (up to 60%)
- Worse conditions in Southern Europe, e.g. water availability, hydropower, summer tourism and crop productivity
- Lower summer precipitation in Central and Eastern Europe. Lower forestry productivity
- Mixed effects in Northern Europe e.g. more floods, but lower heating costs, higher crop yield, higher forest growth
- Sufficient opportunities for adaptation to climate change

policy towards climate change

Dutch Kyoto target for 2008-2012 is −6%.
How does the Netherlands reach this target?

- -1,5% in domestic emissions through sharp reduction
  in emissions of non-CO_2 gases and decoupling CO_2
  from economic growth of about 2% per year in 1990 –
  2012

- About 75 mln tCO_2 eq. emission reduction via CDM
  and JI

Other aspects

Broader climate strategy:

- European policies, incl. ETS
- UNFCCC Post-2012 negotations
- Adaption: raising dikes, flood-reservoirs, spatial planning

- Extra argument: energy security
- Opportunities for market and innovation
- Motivated population, interested business-sector
- Nuclear energy back on national agenda

Reversing the trend

Projection 2020 current policy

AMBITIE IN ONTWIKKELING BROEIKASGASEMISSIE

Projectie 2010 huidig beleid

Doel van het kabinet

Broeikasgasemissie in megatonnen

▬♦▬ Werkelijke (geprojecteerde) broeikasgasemissie

De emissiereductie-doelstelling van het kabinet houdt een trendbreuk in (bron: ECN/MNP)

Policy instruments

- Market incentives
- Standards
- Temporary financial incentives
- Climate and energy diplomacy
- Innovation

The   Approach

- Five sectors: Building Sector, Energy, Industry, Traffic/transport, Agriculture/horticulture
- Separate sector, separate targets
- Agreements with sectors on how to reach the ambitious targets

Sectors (examples)

- **Building Sector**: energy efficiency improvement in around 500,000 buildings of 20-30% till 2011
- **Energy sector:** European Emissions Trading Scheme, subsidy scheme for renewable energy (SDE), 2 demonstration pilots Carbon Capture Storage (CCS)
- **Industry sector:** tightening energy savings covenants, energy saving programs with industrial sectors with goals up to 50% in 2030; ETS
- **Traffic and transport:** tax measures, bio fuels, EU norms, demonstrations of hydrogen-fuelled vehicles
- **Agriculture and horticulture:** heat & cold storage, ETS in 2011, energy neutral greenhouses in 2020, agricultural sector supplies renewable resources for non-food applications

Philosophy

- Instruments and allocation of means are based on sound economics:

  - European approach where possible
  - Uncertainties urge a balanced development to 2020
  - Priority on energy efficiency
  - Prioritizing the budget through a cost-effectiveness check by the Dutch Energy Research Centre (ECN) €/ton CO_2

Finance

- Existing yearly State budget Euro 1.4 billion

- Additional means Euro 0.125 - 0.5 billion in 2011

- 25% of extra budget dedicated to innovation

Tensions

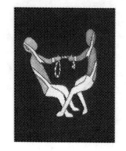

- EU goals vs. national goals

- High ambitions vs. level playing field

- Dependence on EU policy

- Climate policy vs. other national priorities

EU climate and energy package

Proposals presented by EU Commission in January 2008:
- Revised Directive on European Emission Trade System
- Decision on burden-sharing among Member States with regard to CO_2 emission reduction targets
- Directive on renewable energy (incl. biofuels)
- Directive and Communication on CO_2-Capture- and Storage (CCS)

Proposed EU GHG-emissions reduction target

- 20% reduction on 1990 levels

- 1990-2005: 6% reduction achieved

- So, 14% reduction on 2005 levels:

  - ETS industries: 21% reduction on 2005 levels,
    CO2-permit-auction for 2013-2020

  - Non-ETS industries: 10% reduction on 2005 levels,
    subdivided over 27 EU Member States

- Discussions between Member States ongoing

EU climate change diplomacy in Korea

- EC Directive encouraging MS to co-ordinate activities in 3rd countries
- Korea important country for EU to achieve global climate change objectives:
  - 13th largest world economy, 10th largest energy consumer
  - 9th largest emitter of GHG, one of highest per-capita emitters of CO2
  - Bridging gap between developed and developing countries
- Informal Climate Change Attaches Network
- Hopes for Korea: ambitious Copenhagen action, including binding reduction targets, ETS-compatible system, renewable energy and CCS for mitigation

Double Skin Facade

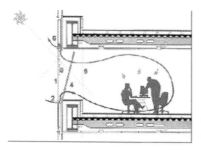

Doble Skin Facade
1. Outer Skin
2. Outer Skin
3. Vertical Blind

5. Inner Skin
6. Outer Skin

An example of Double Skin Buidling

Germany

Dusseldorf. Germany
1998
Story High Cavity
Corridor Type
Natural Ventilation

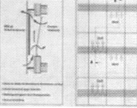

Bibliography

Anderson, Will. (2006). *Diary of an Eco-Builder*, Green Books

Anderson , Curtis and Judy Anderson. (2004). *Electric and Hybrid Cars*: A History, McFarland & Company

Asafu-Adjaye, John. (2005). *Environmental Economics for Non-Economists: Techniques and Policies for Sustainable Development*, World Scientific Publishing Company, Singapore

Baily, Martin Neil and Robert Z. Lawrence, 2001. "Do We Have a New E-conomy?" American Economic Review, 91(2), 308-312

Bassam, N. El (ed). (1998). *Energy Plant Species*: Their Use and Impact on Environment and Technology, Earthscan

Bayon, Ricardo, Amanda Hawn and Katherine Hamilton (eds) (2007). *Voluntary Carbon Markets*: An International Business Guide to What They Are and How They Work, Earthscan

Benson, Charlotte and Edward Clay. (2004). *Understanding the Economic and Financial Impacts of Natural Disasters*, The World Bank

Boken, Vijendra K., Arthur P. Cracknell, Ronald L. Heathcote. (2005). *Monitoring and Predicting Agricultural Drought:* A Global Study, Oxford University Press, USA

Bond ,Patrick, Rehana Dada and Graham Erion ed. (2009). *Climate change, carbon trading and civil society* : negative returns on South African investments. Scotsville, South Africa : University of KwaZulu-Natal Press

Boschert, Sherry. (2006). *Plug-In Hybrids*: The Cars that Will Recharge America, New Society

Botterill, Linda & Donald Wilhite (eds). (2006). *From Disaster Response to Risk Management*: Australia's National Drought Policy, Springer

Boyle, Godfrey. (2004). *Renewable Energy*, Oxford University Press

Bradford, Travis. (2006). *Solar Revolution*: The Economic Transformation of the Global Energy Industry, MIT Press

Brower, David Ross. (2000). *Let the Mountains Talk, Let the Rivers Run*: A Call to Those Who Would Save the Earth. Gabriola Island, B.C.: New Society Publishers

Burroughs, William James. (2001). *Climate Change: A Multidisciplinary Approach*. NY: Cambridge University Press

Burt, Christopher. (2007). *Extreme Weather*: A Guide and Record Book, W. W. Norton & Co

Carraro, C. (ed.). (2000). *Efficiency and Equity of Climate Change Policy*, Springer

Chandler, Alfred, 1990. Scale and Scope: The Dynamics of Industrial Capitalism. Cambridge, MA: Harvard University Press

Chandrasekharam, D. (2002). *Geothermal Energy Resources for Developing Countries,* Taylor & Francis

Chang, C.P. (2004). *East Asian Monsoon* (World Scientific Series on Meteorology of East Asia), World Scientific Publishing

Chiras, Daniel D. (2006). *The Homeowner's Guide to Renewable Energy*: Achieving Energy Independence Through Solar, Wind, Biomass And Hydropower, New Society Publishers

Cipiti, Ben. (2007). The Energy Construct: Achieving a Clean, Domestic and Economical Energy Future, Booksurge Publishing

Claussen, Eileen. (2001). Vicky Arroyo Cochran, Debra P. Davis, eds. Climate Change: Science, Strategies, & Solutions. Boston: Brill

Cowie, Jonathan. (2007). Climate Change: Biological and Human Aspects, Cambridge University Press

Crick, H.Q.P. and Sparks, T.H. (1999). Climate change related to egg laying trends, *Nature* 399, 423-424.

Cudahy, Brian J. (2006). Box boats: How container ships changed the world. New York: Fordham University Press

Daniels, Ronald J., Donald F. Kett, Howard Kunreuther. (2006). On Risk and Disaster: Lessons from Hurricane Katrina, University of Pennsylvania

Daufresne, M., M. C. Roger, H. Capra and N. Lamouroux. (2004). Long-term change within the invertebrate and fish communities of the upper Rhône river: effects of climatic factors, *Global Change Biology*, 10, 124-140.

Dauncey, Guy. (2001). Stormy Weather: 101 Solutions to Global Climate Change. Gabriola Island, B.C.: New Society Publishers

Davidson, I. C. and M. S. Hazlewood. (2005). *Effect of climate change on salmon fisheries,* Environment Agency, UK, Science Report W2-047/SR.

Dawson, Brian and Matt Spannagle. (2009). The complete guide to climate change London; New York : Routledge

Delécolle, R. (2000). Modelling climate change impacts on winter wheat in the Paris basin, in Downing *et al.*

Dessler, Andrew Emory. (2006). The Science and Politics of Global Climate Change: A Guide to the Debate. NY: Cambridge University Press

Dessler, Andrew E. and Edward A. Parson. (2005). The Science and Politics of Global Climate Change: A Guide to the Debate, Cambridge University Press

Dickson, Mary H. and Mario Fanelli. (2005). Geothermal Energy: Utilisation and Technology, Earthscan

Dirnböck, T., S. Dullinger and G. Grabherr. (2003). A regional impact assessment of climate and land-use change on alpine vegetation, *Journal of Biogeography*, 30, 401-417.

Dobson, F. (1981) *Lichens: An Illustrated Guide,* Kingprint Ltd, Surrey.

Doering, Otto C., J. C. Randolph, Jane Southworth et al (eds.). (2002). Effects of Climate Change and Variability on Agricultural Production Systems, Springer

Dow, Kirstin. (2006). The Atlas of Climate Change: Mapping the World's Greatest Challenge. Berkeley: University of California Press

Downie, David Leonard. (2009). Climate change: a reference handbook / David L. Downie, Kate Brash, and Catherine Vaughan. Santa Barbara, Calif. : ABC-CLIO

Downing, T. E., P. A. Harrison, R. E. Butterfield and K. G. Lonsdale. (2000). *Climate Change, Climatic Variability and Agriculture in Europe: An Integrated Assessment,* Environmental Change Institute, Oxford

Drake, Frances. (2000). *Global Warming:* The Science of Climate Change. NY: Oxford University Press

Drapcho, Caye, John Nghiem and Terry Walker. (2008). *Biofuels Engineering Process Technology,* McGraw-Hill Professional, USA

Drucker, Peter. (2002) *Managing in the Next Society,* New York: Truman Talley Books.

Dryzek, John S. (2005). The *Politics of the Earth: Environmental Discourses.* NY: Oxford University Press

Dunning, John H. (2005) Is Global Capitalism morally defensible?', Political Economy Vol.24, no. 1, pp.135-151.

EEA. (2006). Report no.7. *How much bioenergy can Europe produce without harming the environment?*

Eitzinger, J., M. Stastná, Z. Zalud and M. Dubrovsky. (2003). 'A simulation study of the effect of soil water balance and water stress on winter wheat production under different climate change scenarios', *Agricultural Water Management*, 61, 195-217.

Elliot, David. (2003). *A Solar World* (Schumacher Briefing no.10), Green Books

Elsayed, M. A., R. Matthews and N. D. Mortimer. (2003). *Carbon and Energy Balances for a Range of Biofuels Options*, Sheffield Hallam University.

Emanuel, Kerry. (2007). *What We Know About Climate Change*, The MIT Press

Eno Transportation Foundation. (2002). *Global Climate Change and Transportation:* Coming to Terms, Eno Transportation Foundation

Est, Rinie Van. (2000). *Winds of Change: A Comparative Study on the Politics of Wind Energy Innovation in California and Denmark*, International Books, Netherlands

Esty, Daniel C. and Andrew S. Winston. (2006). *Green to Gold*: How Smart Companies Use Environmental Strategy to Innovate, Create Value and Build Competitive Advantage, Yale University Press

European Environment Agency. (2004). *Impacts of Europe's Changing Climate:* an indicator based assessment, Copenhagen.

Ewert, F., M. D. A. Rounsevell, I. Reginster, M. J. Metzger and R. Leemans, (2005). 'Future scenarios of European agricultural land use, Estimating changes in crop production', *Agriculture, Ecosystems and Environment*, 107, 101-116.

Firor, John. (2002). *The Crowded Greenhouse:* Population, Climate Change, and Creating a Sustainable World. New Haven: Yale University Press

Flannery, Tim F. (2005). The Weather Makers: How Man is Changing the Climate and What It Means for Life on Earth. NY: Atlantic Monthly Press

Flannery, Tim. (2006). We Are the Weather Makers: The Story of Global Warming, The Text Publishing Company, Australia

Fleming, James Rodger. (2004). *Historical Perspectives on Climate Change*, Oxford University Press

Friedland, K. D. (1998). 'Ocean climate influences on critical Atlantic Salmon *(Salmo salar* L.*)* life history events', *Canadian Journal of Fish. Aquat. Sci.,* 55, pp.119-130.

Friedland, K. D., L. P. Hansen, D. A. Dunkley and J. C. Maclean. (2004). 'Linkage between ocean climate, post-smolt growth and survival of Atlantic Salmon *(Salmo salar* L.*)* in the North Sea area', ICES, *Journal of Marine Science,* 57, 419-429.

Fulco Ludwig et al. ed. (2009). Climate change adaptation in the water sector / London ; Sterling, VA : Earthscan

Gelbspan, Ross. (2004). *Boiling Point*: How Politicians, Big Oil and Coal, Journalists and Activists Have Fuelled the Climate Crisis – And What We Can Do to Avert Disaster, Basic Books

Gilpin, Alan. (2006). *Environmental Impact Assessment (EIA):* Cutting Edge Impact for the Twenty-First Century, Cambridge University Press

Gipe, Paul. (1999). Wind Energy Basics: A Guide to Small and Micro Wind Systems, Chelsea Green Publishing Co.

Glantz, Michael H. (2003). Climate Affairs: A Primer, Island Press

Gleick, Peter. (2001). "Climate Change in a Warming World." *California Water News* 4 Jan. 2001.

Good, P., L. Barring, C. Giannakopoulos, T. Holt and J. P. Palutikof. (2004). Non-linear regional relationships between climate extremes and annual mean temperatures in model projections for 1961-2099 over Europe. MICE project special issue, *Climate Dynamics*

Gottfried, M., H. Pauli, K. Reiter and G. Grabherr, (1999). A fine-scaled predictive model for changes in species distribution patterns of high mountain plants induced by climate warming, *Diversity and Distributions,* 5, 241-251.

Grabherr, G., M. Gottfried and H. Pauli, (1994). Climate effects on mountain plants, *Nature*, 369, p.448.

Grabherr, G., M. Gottfried and H. Pauli. (2001). Long-term monitoring of mountain peaks in the Alps, *Tasks for Vegetation Science*, 35, pp.153-177.

Grace, J., F. Berninger and L. Nagy, (2002). Impacts of climate change on the tree line, *Annals of Botany*, 90, 537-544.

Griffiths, H. (2005). *The Carbon Balance of Forest Biomes*, Taylor & Francis

Grubb, M. and J. Wilde. (2004). *The European Emissions Trading Scheme*: Implications for Industrial Competitiveness, The Carbon Trust (http://www.carbontrust.co.uk/default.ct)

Grubb, M. (2005). *The Climate Change Challenge*. The Carbon Trust, 2005. (http://www.carbontrust.co.uk/default.ct)

Guha, Ramachandra. (2006). *How Much Should a Person Consume?* Environmentalism in India and the United States, University of California Press

Gupta, Joyeeta. (2006). *Our Simmering Planet*: What to Do about Global Warming, Zed Books

Hansen, Niles. (2002) 'Dynamic Externalities and Spatial Innovation Diffusion: Implications for Peripheral Regions', *International Journal of Technology, Policy, Management*, Vol.2, no.3, pp.260-271.

Harrison, P. A., R. E. Butterfield and T. E. Downing, (eds) (1995). *Climate Change and Agriculture in Europe: Assessment of Impacts and Adaptations*, Environmental Change Unit, Oxford.

Hart, Stuart L. (2007). *Capitalism at the Crossroads*, Wharton School Publishing

Haurie, Alain and Laurent Viguier (eds). (2006). *The Coupling of Climate and Economic Dynamics*: Essays on Integrated Assessment, Springer

Hatton, Timothy J. (2006). *Global Migration and the World Economy*: Two Countries of Policy and Performance, The MIT Press

Hawken, Paul. (2007). *Blessed Unrest*: How the Largest Movement in the World Came Into Being. NY: Viking

Hawken, Paul., Amory Lovins and Hunter Lovins. (1999). *Natural Capitalism:* Creating the Next Industrial Revolution, Little, Brown & Co

Hayden, Howard. (2005). *The Solar Fraud*: Why Solar Energy Won't Run the World, Vales Lake Publishing

Heller, Peter S. (2003). *Who Will Pay? Coping with Aging Societies, Climate Change and Other Long-Term Fiscal Challenges*, International Monetary Fund,

Henson, Robert. (2006). *The Rough Guide to Climate Change*. The Symptoms. The Science. The Solutions, Rough Guides

Holden, N. M. and A. J. Brereton, (2002). An assessment of the potential impact of climate change on grass yield in Ireland over the next 100 years, *Irish Journal of Agricultural and Food Research*, 41, 213-226.

Holden, N. M., A. J. Brereton, R. Fealy and J. Sweeney, (2003). Possible change in Irish climate and its impact on barley and potato yields, *Agricultural and Forest Meteorology*, 116.

Hulme, M., G.J. Jenkins, X. Lu, J.R. Turnpenny, T.D. Mitchell, R.G. Jones, J. Lowe, J.M. Murphy, D. Hassell, P. Boorman, R. McDonald and S. Hill, (2002) *Climate Change Scenarios for the United Kingdom: The UKCIP02 Scientific Report*, Tyndall Centre for Climate Change Research, School of Environmental Sciences, University of East Anglia, Norwich, UK.

Hyde, Richard(ed). (2007). *Bioclimatic Housing*: Innovative Designs for Warm Climates, Earthscan

Imhoff, Daniel and Roberto Carra. (2005). *Paper or Plastic*: Searching for Solutions to an Overpackaged World, Sierra Book Clubs

IPCC 2001a, *Climate Change 2001: The Scientific Basis,* Cambridge University Press.

IPCC 2001b, *Climate Change 2001: Impacts, Adaptation, and Vulnerability,* Cambridge University Press.

IPCC 2001c, *Climate Change 2001: Mitigation,* Cambridge University Press.

Jaccard, Mark. (2005). *Sustainable Fossil Fuels*: The Unusual Suspect in the Quest for Clean and Enduring Energy, Cambridge University Press

Jackson, Tim (ed.). (2001). *Mitigating Climate Change*: Flexibility Mechanisms, Elsevier

Johansen, Bruce E. (2002). *The Global Warming Desk Reference.* Westport, CT: Greenwood Press

Johansen, Bruce E. (2006). *Global Warming in the Twenty-First Century.* Westport, CT: Praeger Publishers

Karling, Horace M., ed. (2001). *Global Climate Change.* NY: Nova Science Publishers

Kim, Junmo (2008). "An Exit for IT industry?: Market Saturation and the Convergence of Ubiquitous technology and manufacturing & service sectors" *International Journal of Technology Management*

Kim, Jnmo (2006b) 'Determinants for sustainability: CO2 emission in time-series analysis' *International Journal of Technology, Policy and Management* Volume 6, Number 4.

Kim, Junmo (2006a) 'Infra Structure of the Digital Economy' *Technological Forecasting & Social Change* Vol 73. NO.4.

Kim, Junmo (2006b) "Will Technology Fusion induce the Paradigm Change of University Education?" *International Journal of Technology Management*

Kim, Junmo (2005a) "Are Industries destined toward the Productivity Paradox,?"
International Journal of Technology Management (IJTM) Vol. 29. No. 3/4.
Inderscience

Kim, Junmo (2005b) *Globalization and Industrial Development.* iUniverse: New York.

Kim, Junmo (2002c) <u>*Commercializing Government's data services*</u>: Its feasibility and limitations A Korea Institute of Public Admin. (KIPA) publication (in Korean)

Kim, Junmo (2002d), <u>*A Cost Benefit Analysis of the recovery of aeronautical meteorology*</u> A contract research by the Korea Meteorological Agency

Kim, Junmo (2001a) "Economic Integration of Major Industrialized Areas" *Technological Forecasting & Social Change.* Vol 67. no.2-3. June 2001.

Kim, Junmo (2000b), <u>*A Study on Cost Recovery of Aeronautical Weather Services*</u> A contract research by the Korea Meteorological Agency

Kirsch, David. (2000). *The Electric vehicle and the Burden of History*, Rutgers University Press

Klinenberg, Eric. (2002). *Heat Wave*: A Social Autopsy of Disaster in Chicago, University of Chicago Press

Kojima, Toshinori. (1998). *The Carbon Dioxide Problem*: Integrated Energy and Environmental Policies for the 21st Century, CRC

Kurokawa, Kosuke (ed). (2007). *Energy from the Desert*: Practical Proposals for Very Large Scale Photovoltaic Systems, Earthscan

Kushner, James A. (2009). *Global climate change and the road to extinction : the legal and plannng response* Durham, N.C. : Carolina Academic Press

Labatt, Sonia and Rodney R. White (eds). (2007). *Carbon Finance*: The Financial Implications of Climate Change, John Wiley & Sons

Langley, Billy and Dan Curtis.(2004). *Going with the Flow*: Small Scale Water Power, Centre for Alternative Technology Publications

Larminie, James and John Lowry. (2003). *Electric Vehicle Technology Explained*, Wiley

Lecocq, Frank. (2005). *State and Trends of the Carbon Market 2004* (World Bank Working Paper), World Bank

Leroux, Marcel. (2005). Global *Warming: Myth or Reality?* The Erring Ways of Climatology, Springer Praxis Books

Lovelock, James. (2006). *The Revenge of Gaia*: Earth's Climate in Crisis and the Fate of Humanity. NY: Basic Books

Lynas, Mark. (2007*). Carbon Counter*, Collins

Maddison, David. (2001).*The Amenity Value of the Global Climate*, Earthscan

Mannion, A. M.(2006). *Carbon and its Domestication*, Kluwer Academic Publishers

McCaffrey, Paul, ed. (2006). Global Climate Change. NY: H.W. Wilson Company

Maracchi, Gianpiero, Oleg Sirotenko and Marco Bindi. (2005).' Impacts of present and future climate change variability on agriculture and forestry in the temperate regions', *Climatic Change*, 70, pp.117-135.

Markandya, Anil and Kirsten Halsnaes (eds). (2002). *Climate Change and Sustainable Development: Prospects for Developing Countries*, Earthscan

Mathew, Sathyajith. (2006). *Wind Energy*: Fundamentals, Resource Analysis and Economics, Springer

McKibben, Bill. (2007). *Deep Economy*: The Wealth of Communities and the Durable Future, Times Books

Milani, Brian. (2000). *Designing the Green Economy*, Rowman & Littlefirld Publishers

Moser, Susanne C., Lisa Dilling, eds. (2007). *Creating a Climate for Change*: Communicating Climate Change and Facilitating Social Change. NY: Cambridge University Press

Nordhaus, William D. and Joseph Boyer. (2003). *Warming the World*: Economic Models of Global Warming, MIT Press

Nordhaus, William D. (1998). *Economics and Policy Issues in Climate Change*, RFF Press

Oldfield, Frank. (2005). *Environmental Change*: Key Issues and Alternative Perspectives. NY: Cambridge University Press

Olsen, Mancur. (1962). The Logic of Collective Action.

Parry, M. (ed.) (2000). *Assessment of Potential Effects and Adaptations for Climate Change in Europe: The Europe ACACIA Project*. Jackson environment Institute, University of East Anglia, Norwich and European Commission.

Parry, M.L., J.E. Hossell, R.G.H. Bunce, P.J. Jones, T. Rehman, R.B. Tranter, J.S. Marsh, C. Rosenzweig and G. Fischer (1996a) Global and regional land use responses to climate change. In: B.H. Walker and W.L. Steffen (Eds.) *Global change and terrestrial ecosystems,* No.2 IGBP Book Series, Cambridge University Press, pp.466-483.

Pearce, Fred. (2007).*With Speed and Violence*: Why Scientists Fear Tipping Points in Climate Change. Boston: Beacon Press

Penman, J. *et al.* (2004). *Good Practice Guidance for Land Use, Land-Use Change and Forestry.* (IPCC).

Pernick, Ron and Clint Wilder. (2007). *The Clean Tech Revolution*: The Next Big Growth and Investment Opportunity, Collins

Pfeiffer, Dale Allen. (2006). *Eating Fossil Fuels*: Oil, Food and the Coming Crisis in Agriculture, New Society Publishers

Philander, George. (2000). *Is the Temperature Rising*? The Uncertain Science of Global Warming, Princeton University Press

Pringle, Laurence P. (2001).*Global Warming:* The Threat of Earth's Changing Climate. NY: SeaStar Books

Rappaport, Ann. and Sarah Hammond Creighton.(2007). *Degrees That Matter:* Climate Change and Industrial Environments, The MIT Press

Ridzon, Leonard. (1994).*The Carbon Cycle*, Acres

Roaf, Sue, David Crichton, Fergus Nicol.(2005). *Adapting Buildings and Cities for Climate Change:* A 21st century Survival Guide, Architectural Press, Elsevier

Roberts, Paul. (2004).*The End of Oil:* The Decline of the Petroleum Economy and the Rise of a New Energy Order, Bloomsbury

Rogers, Heather. (2006).*Gone Tomorrow:* The Hidden Life of Garbage, The New Press

Romm, Joseph J. (2007). *Hell and High Water:* Global Warming- The Solution and the Politics- And What We Should Do. NY: William Morrow

Ruth, Matthias. (2006). *Smart Growth and Climate Change:* Regional Development, Infrastructure, and Adaptation (New Horizons in Regional Science Series), Edward Elgar Publishing

Ryn, Sim van der. (2000). *The Toilet Papers:* Recycling Waste and Conserving Water, Chelsea Green Pub Co, 2nd revised edition

Sakong, Il editor (1987) *Macroeconomic Policy and Industrial Development Issues* Seoul, KDI

Salomon, Thierry, Stephane Bedel. (2003). *The Energy Saving House* (New Futures), Centre for Alternative Technology Publications

Scheer, Hermann. (2006). *Energy Autonomy:* The Economic, Social and Technological Case for Renewable Energy, Earthscan

Scheer, Hermann. (2004). *The Solar Economy*, Earthscan Publications

Schor, Juliet B., Betsy Taylor, eds. (2002). *Sustainable Planet:* Solutions for the Twenty-First Century. Boston: Beacon Press

Seth, Andrew, Geoffrey Randall. (2005). *Supermarket Wars*: The Future of Global Food Retailing, Palgrave Macmillan

Sinnott-Armstrong, Walter, Richard Howarth, eds. (2005). *Perspectives on Climate Change*: Science, Economics, Politics, Ethics. San Diego, CA: Elsevier

Smith, Peter F.. (2004). *Eco-Refurbishment*: A Guide to Saving and Producing Energy in the Home, Architectural Press, Elsevier

Smith, Peter. (2001). *Architecture in a Climate of Change*, Architectural Press

Snow, T. P. (2007).*Carbon: The Sixth Element*, Princeton University Press

Soon, Willie. (2001). Global *Warming: A Guide to the Science*. Vancouver: Fraser Institute, Centre for Studies in Risk and Regulation

Sperling, Daniel and James Cannon. (2006). *Driving Climate Change*: Cutting carbon from Transportation, Academic Press Inc

Sperling, Daniel Sperling and James Cannon. (2004). *The Hydrogen Energy Transition:* Cutting Carbon from Transportation, Academic Press Inc

Strong, Stephen J. (1993*)*. *The Solar Electric House*: Energy for the Environmentally-Responsive, Energy-Independent House, Chelsea Green Publishing Co, 3rd revised edition

Swingland, Ian. (2003). *Capturing Carbon and Conserving Biodiversity*: The Market Approach, Earthscan Publications

Tang, Kenny (ed). (2005). *The Finance of Climate Change:* A Guide for Governments, Corporations and Investors, Risk Books

Tang, Kenny and Ruth Yeoh. (2007). *Cut Carbon, Grow Profits*: Business Strategies for Managing Climate Change and Sustainability, Middlesex University Press

Tanaka, Shelley. (2006). *Climate Change*. Berkley: Groundwood Books

Tietenberg, T. H.. (2006). *Emissions Trading:* Principles and Practice, RFF Press, 2nd edition

Tim Flannery. (2005). *The Weather Makers:* The History and Future Impact of Climate Change, Penguin, Allen Lane

Toman, Michael A. (ed). (2001).*Climate Change Economics and Policy:* An RFF Anthology, RFF Press

Trabalka, John R. (2005).*Atmospheric Carbon Dioxide and the Global Carbon Cycle,* University Press of the Pacific

Tuba, Zoltan. (ed). (2005*). Ecological Responses and Adaptations of Crops to Rising Atmospheric Carbon Dioxide,* Food Products Press

UK Forestry Commission. (2003). *Forestry Statistics*

Underwood, C.P., Francis Yik. (2004). *Modelling Methods for Energy in Buildings,* Blackwell Publishing
Vernon R (1966) International investment and international trade in the product cycle. *Quarterly Journal of Economics* 80.2: 190-207

Vernon R (1979) The Product Cycle Hypothesis in a New International Environment. *Oxford Bulletin of Economics and Statistics* 41.4: 255-26

Walz, Rainer. (2009). *The economics of climate change policies* : macroeconomic effects, structural adjustments and technological change . Heidelberg : Physica-Verlag

Wigley, T. M. L. and D. S. Schimel. (2005). *The Carbon Cycle,* Cambridge University Press, new edition

Wizelius, Tore. (2007).*Developing Wind Power Projects:* Theory and Practice, Earthscan Publications

Wittwer, Sylvan H. (1995). *Food, Climate, and Carbon Dioxide,* CRC

Weiss, Anthony. (2007).*The Global Food Economy,* Zed Books

World Bank, (2008). *International trade and climate change* : economic, legal, and institutional perspectives. Washington D.C..

Yamin, Farhana. (2005). *Climate Change and Carbon Markets*: A Handbook of Emissions Reduction Mechanisms, Earthscan Publications

Yannas, Simos, Evyatar Erell and Jose Luis Molina. (2005). *Roof Cooling Techniques: A Design Handbook*, Earthscan

Zillman, J.W.(1999). "The National Meteorological Service", *World Meteorological Organization Bulletin*, 48, (2), 129-159.

Index